Environment and Skin

Jean Krutmann · Hans F. Merk

Editors

Environment and Skin

Editors
Jean Krutmann
IUF – Leibniz Research Institute for
Environmental Medicine
Düsseldorf
Germany

Hans F. Merk
Universitätsklinikum Aachen Klinik für
Dermatologie u. Allergologie
Aachen
Germany

ISBN 978-3-319-43100-0   ISBN 978-3-319-43102-4   (eBook)
https://doi.org/10.1007/978-3-319-43102-4

Library of Congress Control Number: 2017955685

Printed on acid-free paper

This Springer imprint is published by Springer Nature
The registered company is Springer International Publishing AG
The registered company address is: Gewerbestrasse 11, 6330 Cham, Switzerland

Contents

Part I   Concepts and Methods

1 **Barrier Skin**.. 3
   Hans F. Merk

2 **Combined, Simultaneous Exposure to Radiation Within
   and Beyond the UV Spectrum: A Novel Approach to Better
   Understand Skin Damage by Natural Sunlight**.............. 11
   Jean Krutmann, Kevin Sondenheimer, Susanne Grether-Beck,
   and Thomas Haarmann-Stemmann

3 **The Impact of Climate Change on Skin and
   Skin-Related Disease**.................................... 17
   Louise K. Andersen

4 **Modern Skin Toxicity Testing Strategies**............... 27
   Susanne N. Kolle, Wera Teubner, and Robert Landsiedel

Part II   Environmental Threats to the Skin

5 **Contact Allergy**....................................... 43
   Stefan F. Martin

6 **Contact Urticaria and Contact Urticaria Syndrome**.......... 51
   Hans F. Merk

7 **Risk Assessment for Contact Allergens**.................... 57
   David A. Basketter

8 **UV and Skin: Photocarcinogenesis**........................ 67
   Allen S.W. Oak, Mohammad Athar, Nabiha Yusuf,
   and Craig A. Elmets

9 **Ambient Particulate Matter and Skin**.................... 105
   Andrea Vierkötter, Jean Krutmann, and Tamara Schikowski

10 **POPs and Skin**... 113
   M.M. Leijs, Janna G. Koppe, T. Kraus, J.M. Baron,
   and H.F. Merk

Part I

Concepts and Methods

Barrier Skin

Hans F. Merk

The skin is a major interface between the body and the environment. In most cases the skin serves as a perfect barrier; however it is dependent on several factors such as the chemical properties of a compound to which the skin is exposed topically, but also its concentration, the contact duration, frequency of exposure, and the exposed surface area influence the amount of penetration which may lead to local reactions including irritation, sensitization, and inflammation but also to penetration and entering the systemic circulation which may result in systemic effects [1, 48].

There are several lines of defenses by the barrier organ skin:

- The chemo-physical barrier of the stratum corneum
- The immunocompetent cells of the epidermis and corium
- The armamentarium of xenobiotica-metabolizing enzymes in epidermal cells such as the keratinocytes as well as antigen-presenting cells, e.g., the Langerhans cells
- The melanocyte-keratinocyte unit and its role in pigmentation and protection against UV radiation

H.F. Merk
Department of Dermatology and Allergology,
RWTH Aachen University, Aachen, Germany

Dohlenfeld 8, 45479 Muelheim adR,
Auf'm Hennekamp 50,
D-40225 Düsseldorf, Germany
e-mail: hans.merk@post.rwth-aachen.de

The main chemo-physical barrier of the skin is located in the outermost layer of the skin, the stratum corneum, which consists out of corneocytes surrounded by lipid regions. The main protein are the keratins; however they need several other proteins for their formation in a way that they can function as a barrier. One of these proteins is filaggrin. Filaggrin is a protein which therefore is essential for a normal barrier formation in stratum corneum, and a diminished expression of this protein is associated with the precipitation of atopic dermatitis and ichthyosis vulgaris if the mutation of filaggrin is homozygous and no active filaggrin results [2]. The water content of the cells in the stratum corneum is with 10–15% much lower than in the keratinocytes of the basal area of the epidermis—the stratum basale—or in other cells of the body with about 75–85%. Therefore there is a gradient of the concentration of water in the epidermis resulting in a water loss which is much higher than the water loss by sweating at room temperature up to 29 °C. Under these conditions, the total amount is about 500 mL/day with a diffusion gradient of 0.5–1.0 mg cm^{-2} h^{-1}. This gradient is highly controlled by the stratum corneum, and the transepidermal water loss (TEWL) can be severalfold enhanced if the stratum corneum is removed. Therefore the TEWL correlates well with the barrier function of the skin [3]. The pH value of stratum corneum in healthy skin is around 4.6–5.6; that means it is acidic. This acidity is necessary for a normal ceramide and lipid

© Springer International Publishing Switzerland 2018
J. Krutmann, H.F. Merk (eds.), *Environment and Skin*,
https://doi.org/10.1007/978-3-319-43102-4_1

production, because it is a prerequisite for a normal function of enzymes such as beta-glucocerebrosidase and sphingomyelinases [4]. The lipid domains are quite important for the skin barrier function. They protect against most xenobiotica to which the skin might be exposed. However there are some exceptions such as compounds with a high lipophilicity, because they are able to permeate best along these lipid domains. They consist of long-chain ceramides, free fatty acids, and cholesterol as main lipid classes. However their organization differs from that of other biological membranes. In stratum corneum, two lamellar phases are present with repeat distances of approximately 6 and 13 nm. Moreover the lipids in the lamellar phases form predominantly crystalline lateral phases, but most probably a subpopulation of lipids forms a liquid phase [5]. Further factors which influence the absorption of xenobiotica include their concentrations in a carrier, the contact duration, and the exposed surface area as well as the molecular weight and irritancy [4]. In particular environmental, occupational, or consumer skin exposure to xenobiotica including pharmaceutical products or cosmetics can result in percutaneous absorption and penetration. Also interactions between pharmaceutical compounds and cosmetics or sanitizers including natural products and plant extracts can occur [6, 7]. Besides direct effect on the skin including its barrier function, immune-mediated skin effects and systemic effects are of major concern. Occupational exposure to metals, epoxy and acrylic resins, rubber additives, and chemical intermediates but also ingredients in sanitizers can lead to immune-mediated effects such as dermatitis and urticarial [8]. In addition the exposure to complex mixtures, excessive hand washing, hand sanitizers, and high frequency of wet work are factors which may augment absorption and penetration of xenobiotica [8]. Xenobiotica with major local or systemic toxicity have been reviewed recently (Table 1.1) [9].

The stratum corneum is not only a barrier for most xenobiotica but also for physical agents such as UV light. Under its influence the differentiation of keratinocytes is altered in a way that the stratum corneum is enlarged which improves its capability to absorb UV light.

Table 1.1 Xenobiotica with noticeable local or systemic toxicity [7, 9–12]

Herbicides
Paraquat
Monochloroacetic acid
Pesticides/insecticides (organophosphates)[a]
Methyl bromide
Malathion
Strychnine
Amitraz
Pyrethroids
Chlorpyrifos
Diazinon
N,N-Diethyl-m-toluamide
Cutaneous medications
Topical β-blockers
Topical α_2-agonists
Dorzolamide/brinzolamide
Topical prostaglandin analogs
Minoxidil
Nitroglycerin
Resorcinol
Tretinoin
Adapalene
Salicylic acid
Sodium sulfacetamide
Nicotine
Glucocorticoides
Occupational contact
Trichloroethylene
Lindane
Chromium
Engine oil
Hydrofluoric acid
Isocyanates
Plasticizers (most commonly phthalate esters)[b]
Cosmeceuticals and Diversa
Mercury
Phenol
Arsenic

[a]Alone in the USA, >40 organophosphate pesticides are registered, and EPA estimates that in 2007 about 33 million pounds were applied [8]
[b]Phthalate esters are ubiquitous and are used to manufacture building materials, household furnishings, clothing, cosmetics, pharmaceuticals, nutritional supplements, medical devices, dentures, children's toys, glow stick, modeling clay, food packaging, automobiles, lubricants, waxes, cleaning materials, and insecticides [8]

The stratum corneum is a perfectly specialized chemo-physical barrier and an umbrella over the epidermis. But also the epidermis functions as a

Fig. 1.1 Cytochrome P450 (CYP) and influx (OATP) as well as efflux proteins (MDR/MRP) which have been characterized in human keratinocytes [13–19]

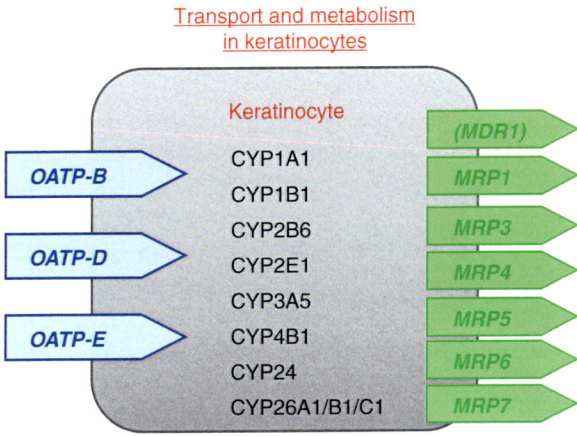

barrier because it works as an immunologic first line of defense and xenobiotica-metabolizing organ. In addition it protects the skin against the damages by UV radiation. The epidermis possesses several highly specialized cells which are connected by several interacting signaling processes with one another. The main compartment are the keratinocytes, which do not only serve as a source for the corneocytes in the stratum corneum compartment, but they contribute to inflammatory processes, e.g., by the production and release of multiple cytokines which modulate inflammatory processes. Keratinocytes play a major role in the synthesis of vitamin D and are able to metabolize xenobiotica by enzymes such as the cytochrome P450 isoenzymes which are inducible by xenobiotica (Fig. 1.1) [21, 44]. About 90% of epidermal cells are keratinocytes; other cells include melanocytes, antigen-presenting dendritic cells—the Langerhans cells and Merkel cells (Table 1.2).

Melanocytes are dendritic cells, which produce melanin, which is transported to connected keratinocytes upon UV radiation, and this melanocyte together with about 30–40 keratinocytes forms a melanocyte-keratinocyte unite, which plays a major role in the UV protection of the skin. The induction of the vitamin D synthesis in keratinocytes by UVB light also augments the expression of filaggrin which improves the barrier formation by the stratum corneum [2].

Further on the skin possesses immunocompetent cells including antigen-presenting dendritic cells and T-lymphocytes which play an important role as an immunologic first line of defense but may also

Table 1.2 Xenobiotica metabolizing enzymes in the skin

Xenobiotica-metabolizing enzymes in the skin
Cytochrome P450
Flavin-dependent monooxygenases (FMO)
Cyclooxygenases (COX1/COX2)
Alcohol dehydrogenase (ADH)
Aldehydedehydrogenase (ALDH)
NAD(P)H:quinone reductase (NQR)
Epoxide hydrolase (EH)
Esterase/amidase
Glutathione S-transferase (GST)
UDP-glucuronosyltransferase (UGT)
Sulfotransferase (SULT)
N-Acetyltransferase (NAT)

lead to sensitization and inflammatory responses to environmental hazards [20]. The epidermal antigen-presenting dendritic cells—the Langerhans cells—are able to sensitize against antigens leading to immunologic tolerance or to hypersensitivity reactions to those compounds. A quite special feature of this sensitization process by the skin is that the skin especially can induce hypersensitivity reactions to small molecular weight compounds as it is the case in allergic contact dermatitis or drug allergy. There are several thousand compounds which are able to cause an allergic contact dermatitis, whereas the number of small molecular weight compounds which, e.g., are able to induce an allergic asthma bronchiale is rather limited to 80–100 substances [17, 21]. In addition in order to induce those allergic reactions of the lung in mice, it is first necessary to sensitize the animals via the skin; there is no primary sensitization, e.g., by inhalation [3, 46].

The pathophysiology of two inflammatory skin diseases which belong to the most often precipitated disorders of the skin—eczema and psoriasis—is closely associated with dysfunctions of the skin barrier. The role of filaggrin in the pathophysiology of atopic dermatitis has been mentioned already. The precipitation of both pathological symptoms—eczema and psoriasis—is dependent on Th1-mediated immunological processes. In the case of eczema formation, one differentiates between atopic dermatitis, which are most often triggered by Th2-dependent allergic reactions, and after a switch of the infiltrating T-lymphocytes to Tc1-type lymphocytes, eczema is precipitated as in the case of allergic contact dermatitis which is triggered and precipitated by Tc1-type lymphocytes [28, 33]. Atopic dermatitis is quite often triggered by protein antigens originating from, e.g., house dust mite, food allergens, and others, whereas allergic contact dermatitis is triggered by small molecular weight compounds.

For centuries both diseases—eczema and psoriasis—were treated by dermatologists with coal tar preparations, which have as a main cellular target a ligand-dependent transcription factor known as aryl-hydrocarbon receptor (AhR) [23]. It has been considered to be something like a sensor for xenobiotica and has been first discovered as mediator of the toxicity of 2,3,7,8-tetrachlorodibenzo-p-dioxin (TCDD) in humans and other mammals and as regulator of the inducible expression of xenobiotica-metabolizing enzymes including cytochrome P450 (CYP) 1A1, 1A2, and 1B1 [23, 25]. However it ruled out that this protein also has fundamental roles in biology including the barrier formation of the skin [26]. In early studies it has already been shown that topical applied coal tar induces AhR-dependent xenobiotica-metabolizing enzymes such as cytochrome P450-dependent CYP1A1 or 1B1 in the skin, which were measured catalytically as aryl-hydrocarbon-hydroxylase (AHH) activity at this time [27, 45]. More recently it has been shown that the topical application of coal tar and its activation of the AhR results in induction of epidermal differentiation and enhanced filaggrin expression in lesional skin of atopic derma-

titis [28, 29]. Also petrolatum is used since many years in dermatology and cosmetology as a common moisturizer, e.g., for maintenance therapy of atopic dermatitis. Topical application of petrolatum also induced expression of filaggrin and other key barrier differentiation markers such as loricrin, leading to an increased stratum corneum thickness of "normal-appearing" or nonlesional AD skin. This beneficial molecular response of petrolatum in barrier-defective states might be related to trace amounts of AhR inducing polycyclic hydrocarbons which might still be present and might therefore mediate its effect on the barrier formation through the AHR [30]. In addition to these effects of AhR induction on skin barrier formation, it has been shown that also immune responses are modulated by the AhR which influences immunocompetent cells, e.g., different T cell subpopulations including Th17 cells resulting in a reduction of the severity of skin inflammatory responses in general by activation of the AhR [31]. Recent observations suggest that coal tar application to patients who are suffering from chronic inflammatory diseases such as eczema or psoriasis may be dependent on the status of AhR signaling. There are a "canonical signaling" under "healthy" conditions, in which AhR induces the AhR gene battery after binding to ARNT (AHR nuclear translocator), and a "noncanonical signaling" if the AhR is induced over a long period of time such as in chronic inflammatory diseases [23]. Under the condition of the "canonical signaling," the induction of AhR increases the risk for carcinogenesis, skin aging, or hyperpigmentation, whereas an induction of the AhR under the condition of "noncanonical signaling" results in a shift of the AhR from the noncanonical to the canonical AhR signaling which reduces the inflammation and normalizes epidermal differentiation and barrier formation [24]. These recent findings of the role of AhR ligands in coal tar and petrolatum ointments as well as their targeted effect on proteins which play a central role in the pathophysiology of atopic dermatitis such as filaggrin may improve the possibility to screen for active ingredients which one can use in skin diseases with impaired barrier formation and inflammatory reactions [32]. This makes the

AhR to a preferred candidate as target protein for new drugs in skin pharmacology. Recently tapinarof (GSK2894512) has been characterized as an AhR ligand and as a potential candidate for the treatment of atopic dermatitis as well as psoriasis [38].

Table 1.1 summarizes the armamentarium of xenobiotica-metabolizing enzymes of the skin with regard to the most important enzymes which have been detected in the skin and which were reviewed recently [14, 21]. The physiological role of these enzymes is to detoxify xenobiotica leading to their elimination from the body. However it may happen that xenobiotica are chemically activated during this metabolism resulting in their binding to critical molecules of the cells which may even lead to carcinogenesis or sensitization by small molecular weight compounds. The metabolism of xenobiotica is divided in a phase I reaction during which most compounds are oxidized to highly reactive species which are further metabolized in a phase II reaction by epoxide hydrolases and transferases to more water-soluble compounds which can be eliminated. Proteins which enable a phase III of the xenobiotica metabolism are transporter proteins which enable the influx and the efflux of those compounds (Fig. 1.1) [47]. The expression of xenobiotica-metabolizing enzymes including cytochrome P450 isoenzymes is dependent on the level of the differentiation of keratinocytes, e.g., most of the cytochrome P450 1 is present in the basal keratinocytes [13]. In addition some specialized keratinocytes such as the follicular keratinocytes possess a much higher CYP-dependent enzyme activity compared to normal keratinocytes [25, 45]. Recently it has been shown that in particular the antigen-presenting dendritic Langerhans cells possess a high CYP 1B1-dependent activity which is able to metabolize dimethylbenzanthracene (DMBA) to a carcinogenic derivative which then leads to critical mutations in the genes of keratinocytes, whereas the same CYP-dependent activity is not sufficient in keratinocytes to activate DMBA [34]. This observation is consistent with the finding that also other antigen-presenting cells such as monocytes express high amounts of CYP 1B1 mRNA (Fig. 1.1 and 1.2) [35]. On a protein level,

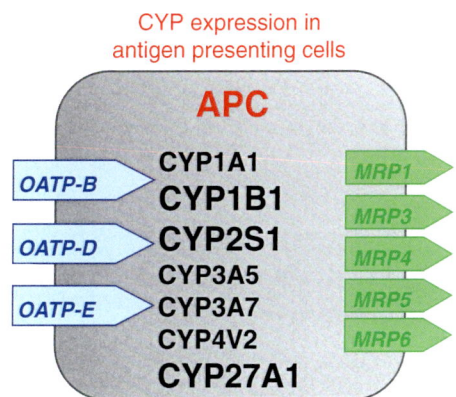

Fig. 1.2 Cytochrome P450 (CYP) and influx (OATP) as well as efflux proteins (MDR/MRP) which have been characterized in antigen-presenting cells [34, 35, 37, 38]

the presence of CYP1A1, 2B6, 2E1, and 3A has been shown. On the mRNA level, the treatment of keratinocytes with dexamethasone resulted in an induction of CYP3A4 and downregulation of CYP3A5 [13] (Table 1.3).

Several phase II enzymes are present in the skin such as isoforms of glutathione-S-transferase, UDP-glucuronosyltransferase, N-acetyltransferases (NATs), and in particular sulfotransferases. NATs are expressed in the human skin and play an important role in the detoxification of hair dyes that contain aromatic amines, which are the main culprit in allergic reactions of hairdressers and which are potentially carcinogenic [39]. NAT-1 mRNA has been detected in human keratinocytes, and a polymorphism of NAT-1 gene may influence the sensitization risk to hair dyes such as p-phenylenediamine [40]. Of increasing interest is the sulfotransferase superfamily of enzymes (SULTs), which have significant activity in keratinocytes and may play a role in the metabolism of topical or systemically applied drugs such as minoxidil or nevirapine [41–43].

Taken together the skin possesses several effective first lines of defenses of the body; however if this barrier function is stressed too much by environmental hazards, serious diseases including irritant or allergic contact dermatitis, atopic dermatitis, photosensitization, urticaria, porphyria, and acne may result, which will be further discussed in other chapters of this book.

Table 1.3 Cutaneous CYP-isoenzymes

CYP isoenzyme	Baron et al. (2008)	Yengi et al. (2003)	Smith et al. (2006)	Wiegand et al. (2014)
1A1	+	+	+	+
1A2	+	−	+	−
1B1	+	+	+	+
2A6/7	+	−	−	+
2B6/7	+	+	−	−
2C9	+	+	−	−
2C18	+	+	+	−
2C19	+	+	−	−
2D6	+	+	−	−
2E1	+	+	−	+
2S1	+	−	−	+
3A4/7	+	+	−	+
3A5	+	+	+	−
4B1	+	−	−	−
4X1	+	−	−	−
19A1	+	−	−	−
26B1	+	−	+	−

Source: Baron JM, Wiederholt T, Heise R, Merk HF, Bickers DR. Expression and function of cytochrome p450-dependent enzymes in human skin cells. Curr Med Chem. 2008;15(22):2258–64; Smith G, Ibbotson SH, Comrie MM, Dawe RS, Bryden A, Ferguson J, Wolf CR. Regulation of cutaneous drug-metabolizing enzymes and cytoprotective gene expression by topical drugs in human skin in vivo. Br J Dermatol. 2006;155(2):275–81; Wiegand C, Hewitt NJ, Merk HF, Reisinger K. Dermal xenobiotic metabolism: a comparison between native human skin, four in vitro skin test systems and a liver system. Skin Pharmacol Physiol. 2014;27(5):263–75. doi: 10.1159/000358272; Yengi LG, Xiang Q, Pan J, Scatina J, Kao J, Ball SE, Fruncillo R, Ferron G, Roland Wolf C. Quantitation of cytochrome P450 mRNA levels in human skin. Anal Biochem. 2003;316(1):103–10

References

1. Roelofzen JH, Aben KK, Oldenhof UT, Coenraads PJ, Alkemade HA, van de Kerkhof PC, van der Valk PG, Kiemeney LA. No increased risk of cancer after coal tar treatment in patients with psoriasis or eczema. J Invest Dermatol. 2010;130(4):953–61.
2. Merk HF, Baron JM, Neis MM, Obrigkeit DH, Karlberg AT. Skin: major target organ of allergic reactions to small molecular weight compounds. Toxicol Appl Pharmacol. 2007;224(3):313–7.
3. Proksch E, Brandner JM, Jensen JM. The skin: an indispensable barrier. Exp Dermatol. 2008;17(12):1063–72.
4. Blickenstaff NR, Coman G, Blattner CM, Andersen R, Maibach HI. Biology of percutaneous penetration. Rev Environ Health. 2014;29(3):145–55.
5. Bouwstra JA, Ponec M. The skin barrier in healthy and diseased state. Biochim Biophys Acta. 2006;1758(12):2080–95.
6. Muhammad F, Wiley J, Riviere JE. Influence of some plant extracts on the transdermal absorption and penetration of marker penetrants. Cutan Ocul Toxicol. 2017;36(1):60–6.
7. Neis MM, Wendel A, Wiederholt T, Marquardt Y, Joussen S, Baron JM, Merk HF. Expression and induction of cytochrome p450 isoenzymes in human skin equivalents. Skin Pharmacol Physiol. 2010;23(1):29–39.
8. Anderson SE, Meade BJ. Potential health effects associated with dermal exposure to occupational chemicals. Environ Health Insights. 2014;8(Suppl 1):51–62.
9. Alikhan FS, Maibach H. Topical absorption and systemic toxicity. Cutan Ocul Toxicol. 2011;30(3):175–86. doi:10.3109/15569527.2011.560914. [Epub 2011 Mar 22]
10. Buchan P, Jamoulle JC. Percutaneous absorption. In: Soter NA, Baden HP, editors. Pathophysiology of dermatologic diseases. New York: McGraw-Hill; 1991. p. 83–90.
11. Lu W, Uetrecht JP. Possible bioactivation pathways of lamotrigine. Drug Metab Dispos. 2007;35:1050–6.
12. van den Bogaard EH, Podolsky MA, Smits JP, Cui X, John C, Gowda K, Desai D, Amin SG, Schalkwijk J, Perdew GH, Glick AB. Genetic and pharmacological analysis identifies a physiological role for the AHR in epidermal differentiation. J Invest Dermatol. 2015;135(5):1320–8.
13. Baron JM, Höller D, Schiffer R, Frankenberg S, Neis M, Merk HF, Jugert FK. Expression of multiple cytochrome p450 enzymes and multidrug resistance-associated transport proteins in human skin keratinocytes. J Invest Dermatol. 2001;116(4):541–8.
14. Baron JM, Wiederholt T, Heise R, Merk HF, Bickers DR. Expression and function of cytochrome p450-dependent enzymes in human skin cells. Curr Med Chem. 2008;15(22):2258–64.
15. Heise R, Mey J, Neis MM, Marquardt Y, Joussen S, Ott H, Wiederholt T, Kurschat P, Megahed M, Bickers DR, Merk HF, Baron JM. Skin retinoid concentrations are modulated by CYP26AI expression restricted to basal keratinocytes in normal human skin and differentiated 3D skin models. J Invest Dermatol. 2006;126(11):2473–80.
16. Heise R, Skazik C, Rodriguez F, Stanzel S, Marquardt Y, Joussen S, Wendel AF, Wosnitza M, Merk HF, Baron JM. Active transport of contact allergens and steroid hormones in epidermal keratinocytes is mediated by multidrug resistance related proteins. J Invest Dermatol. 2010;130(1):305–8.
17. North CM, Ezendam J, Hotchkiss JA, Maier C, Aoyama K, Enoch S, Goetz A, Graham C, Kimber I, Karjalainen A, Pauluhn J, Roggen EL, Selgrade M, Tarlo SM, Chen CL. Developing a framework for assessing chemical respiratory sensitization: a workshop report. Regul Toxicol Pharmacol. 2016;80:295–309.

18. Sebastian K, Detro-Dassen S, Rinis N, Fahrenkamp D, Müller-Newen G, Merk HF, Schmalzing G, Zwadlo-Klarwasser G, Baron JM. Characterization of SLCO5A1/OATP5A1, a solute carrier transport protein with non-classical function. PLoS One. 2013;8(12):e83257. doi:10.1371/journal.pone.0083257.

19. Sharma AM, Novalen M, Tanino T, Uetrecht JP. 12-OH-nevirapine sulfate, formed in the skin, is responsible for nevirapine-induced skin rash. Chem Res Toxicol. 2013;26(5):817–2.

20. van den Bogaard EH, Bergboer JG, Vonk-Bergers M, van Vlijmen-Willems IM, Hato SV, van der Valk PG, Schröder JM, Joosten I, Zeeuwen PL, Schalkwijk J. Coal tar induces AHR-dependent skin barrier repair in atopic dermatitis. J Clin Invest. 2013;123(2):917–27.

21. Oesch F, Fabian E, Guth K, Landsiedel R. Xenobiotic-metabolizing enzymes in the skin of rat, mouse, pig, guinea pig, man, and in human skin models. Arch Toxicol. 2014;88(12):2135–90.

22. Robinson PJ. Prediction: simple risk models and overview of dermal risk assessment. In: Roberts MS, Walters KA, editors. Dermal absorption and toxicity assessment. New York: Marcel Dekker; 1998. p. 203–29.

23. Haarmann-Stemmann T, Esser C, Krutmann J. The Janus-faced role of aryl hydrocarbon receptor signaling in the skin: consequences for prevention and treatment of skin disorders. J Invest Dermatol. 2015;135(11):2572–6.

24. Roelofzen JH, Aben KK, Van de Kerkhof PC, Van der Valk PG, Kiemeney LA. Dermatological exposure to coal tar and bladder cancer risk: a case-control study. Urol Oncol. 2015;33(1):20.e19–22.

25. Merk HF, Sachs B, Baron J. The skin: target organ in immunotoxicology of small-molecular-weight compounds. Skin Pharmacol Appl Skin Physiol. 2001;14(6):419–30.

26. Haas K, Weighardt H, Deenen R, Köhrer K, Clausen B, Zahner S, Boukamp P, Bloch W, Krutmann J, Esser C. Aryl hydrocarbon receptor in keratinocytes is essential for murine skin barrier integrity. J Invest Dermatol. 2016;136(11):2260–9. doi:10.1016/j.jid.2016.06.627.

27. Bickers DR, Kappas A. Human skin aryl hydrocarbon hydroxylase. Induction by coal tar. J Clin Invest. 1978;62(5):1061–8.

28. Weidinger S, Novak N. Atopic dermatitis. Lancet. 2016;387(10023):1109–22.

29. Wester RC, Maibach HI. Animal models for percutaneous absorption. In: Wang RGM, Knaak JB, Maibach HI, editors. Health risk assessment. Boca Raton: CRC Press; 1993. p. 89–103.

30. Czarnowicki T, Malajian D, Khattri S, Correa da Rosa J, Dutt R, Finney R, Dhingra N, Xiangyu P, Xu H, Estrada YD, Zheng X, Gilleaudeau P, Sullivan-Whalen M, Suaréz-Fariñas M, Shemer A, Krueger JG, Guttman-Yassky E. Petrolatum: barrier repair and antimicrobial responses underlying this "inert" moisturizer. J Allergy Clin Immunol. 2016;137(4):1091–102.

31. Di Meglio P, Duarte JH, Ahlfors H, Owens ND, Li Y, Villanova F, Tosi I, Hirota K, Nestle FO, Mrowietz U, Gilchrist MJ, Stockinger B. Activation of the aryl hydrocarbon receptor dampens the severity of inflammatory skin conditions. Immunity. 2014;40(6):989–1001.

32. Bickers DR, Das M, Mukhtar H. Pharmacological modification of epidermal detoxification systems. Br J Dermatol. 1986;115(Suppl 31):9–1.

33. Suwanpradid J, Holcomb ZE, MacLeod AS. Emerging skin T-cell functions in response to environmental insults. J Invest Dermatol. 2016;137(2):288–94. doi:10.1016/j.jid.2016.08.013. pii: S0022-202X(16)32347-8, [Epub ahead of print]

34. Modi BG, Neustadter J, Binda E, Lewis J, Filler RB, Roberts SJ, Kwong BY, Reddy S, Overton JD, Galan A, Tigelaar R, Cai L, Fu P, Shlomchik M, Kaplan DH, Hayday A, Girardi M. Langerhans cells facilitate epithelial DNA damage and squamous cell carcinoma. Science. 2012;335(6064):104–8. doi:10.1126/science.1211600.

35. Baron JM, Zwadlo-Klarwasser G, Jugert F, Hamann W, Rübben A, Mukhtar H, Merk HF. Cytochrome P450 1B1: a major P450 isoenzyme in human blood monocytes and macrophage subsets. Biochem Pharmacol. 1998;56(9):1105–10.

36. Muhammad F, Jaberi-Douraki M, de Sousa DP, Riviere JE. Modulation of chemical dermal absorption by 14 natural products: a quantitative structure permeation analysis of components often found in topical preparations. Cutan Ocul Toxicol. 2016;14:1–16.

37. Skazik C, Heise R, Ott H, Czaja K, Marquardt Y, Merk HF, Baron JM. Active transport of contact allergens in human monocyte-derived dendritic cells is mediated by multidrug resistance related proteins. Arch Biochem Biophys. 2011;508(2):212–6.

38. Smith S, Jayawickreme C, Rickard D, Nicodeme E, Bui T, Simmons C, Coquery C, Nei J, Pryor W, Mayhew D, Raypal D, Creech K, Furst S, Lee J, Wu D, Rastinejad F, Willson T, Viviani F, Morris D, Moore J, Cote-Sierra J. Anti-inflammatory activity of a bacterial small molecule product derives from Aryl hydrocarbon Receptor activation. Exp Dermatol. 2016;25(Suppl. 2):49.

39. Kawakubo Y, Merk HF, Masaoudi TA, Sieben S, Blömeke B. N-acetylation of paraphenylenediamine in human skin and keratinocytes. J Pharmacol Exp Ther. 2000;292(1):150–5.

40. Blömeke B, Brans R, Coenraads PJ, Dickel H, Bruckner T, Hein DW, Heesen M, Merk HF, Kawakubo Y. Para-phenylenediamine and allergic sensitization: risk modification by N-acetyltransferase 1 and 2 genotypes. Br J Dermatol. 2009;161(5):1130–5.

41. McLean WH. Filaggrin failure – from ichthyosis vulgaris to atopic eczema and beyond. Br J Dermatol. 2016;175(Suppl 2):4–7.

42. Sharma AM, Uetrecht J. Bioactivation of drugs in the skin: relationship to cutaneous adverse drug reactions. Drug Metab Rev. 2014;46(1):1–18.

43. Skazik C, Heise R, Bostanci O, Paul N, Denecke B, Joussen S, Kiehl K, Merk HF, Zwadlo-Klarwasser G, Baron JM. Differential expression of influx and efflux transport proteins in human antigen presenting cells. Exp Dermatol. 2008;17(9):739–47.

44. Kappas A, Alvares AP, Bickers DR, Levin W, Conney A. The induction of a carcinogen-metabolizing enzyme in human skin. Trans Am Clin Climatol Assoc. 1973;84:125–31.

45. Merk HF, Mukhtar H, Kaufmann I, Das M, Bickers DR. Human hair follicle benzo[a]pyrene and benzo[a] pyrene 7,8-diol metabolism: effect of exposure to a coal tar-containing shampoo. J Invest Dermatol. 1987;88(1):71–6.

46. Pauluhn J. Development of a respiratory sensitiza-tion/elicitation protocol of toluene diisocyanate (TDI) in Brown Norway rats to derive an elicitation-based occupational exposure level. Toxicology. 2014;319:10–22.

47. Schiffer R, Neis M, Höller D, Rodríguez F, Geier A, Gartung C, Lammert F, Dreuw A, Zwadlo-Klarwasser G, Merk H, Jugert F, Baron JM. Active influx transport is mediated by members of the organic anion trans-porting polypeptide family in human epidermal kera-tinocytes. J Invest Dermatol. 2003;120(2):285–91.

48. Wester RC, Maibach HI. Percutaneous absorption. In: Wang RGM, Knaak JB, Maibach HI, editors. Health risk assessment. Boca Raton: CRC Press; 1993. p. 63–87.

Combined, Simultaneous Exposure to Radiation Within and Beyond the UV Spectrum: A Novel Approach to Better Understand Skin Damage by Natural Sunlight

2

Jean Krutmann, Kevin Sondenheimer, Susanne Grether-Beck, and Thomas Haarmann-Stemmann

2.1 Introduction

One of the major threats to human skin health is natural sunlight. There is no doubt that chronic exposure to ultraviolet B (290–320 nm, UVB) radiation is the major cause for nonmelanoma skin cancer in humans, as reviewed extensively in Chap. 7 of this monograph. It is now also well established that longer UV radiation, i.e., UVA (320–400 nm), is photocarcinogenic as well and, e.g., can cause the formation of cyclobutane pyrimidine dimers in nuclear DNA and thereby exert mutagenic and immunosuppressive effects. An increasing number of studies conducted during the last 15 years, however, suggest that this is not the end of the story but that in fact, wavelengths in natural sunlight beyond the UV spectrum can damage human skin as well. Such studies have mainly focused on two types of nonionizing radiation: (1) the blue light part of visible light (400–495 nm) as well as (2) near-infrared radiation, i.e., IRA (770–1400 nm).

The existing evidence that such radiation types can damage human skin has recently been summarized in a number of state-of-the-art reviews [1, 2]. In this chapter, rather than simply repeating this knowledge, we will provide a brief summary of the major findings coming from such studies and, for more detailed information, refer the interested reader to the respective publications. Instead, we will put forward and discuss the hypothesis that most of our current knowledge as it concerns sunlight-induced health damage to human skin might be biased by the irradiation protocols that were used. Accordingly, in the vast majority of studies, skin responses have been studied which were elicited by either UVB alone or UVA alone or blue light alone or IRA alone. This is in marked contrast to real exposure scenarios, where human skin is exposed to all these wavelengths simultaneously, because they are all present in natural sunlight. We will discuss the existing evidence that signaling responses elicited in skin cells by different wavelengths might interact with and thereby influence each other and that the resulting responses are different from the ones induced by each of the single irradiations. We will also put forward the hypothesis that it is mandatory to study the combined rather than additive effects of different wavelengths within and beyond natural sunlight

J. Krutmann (✉) • K. Sondenheimer
S. Grether-Beck • T. Haarmann-Stemmann
IUF—Leibniz Research Institute for Environmental Medicine, Düsseldorf, Germany
e-mail: Jean.Krutmann@IUF-Duessedorf.de

© Springer International Publishing Switzerland 2018
J. Krutmann, H.F. Merk (eds.), *Environment and Skin*,
https://doi.org/10.1007/978-3-319-43102-4_2

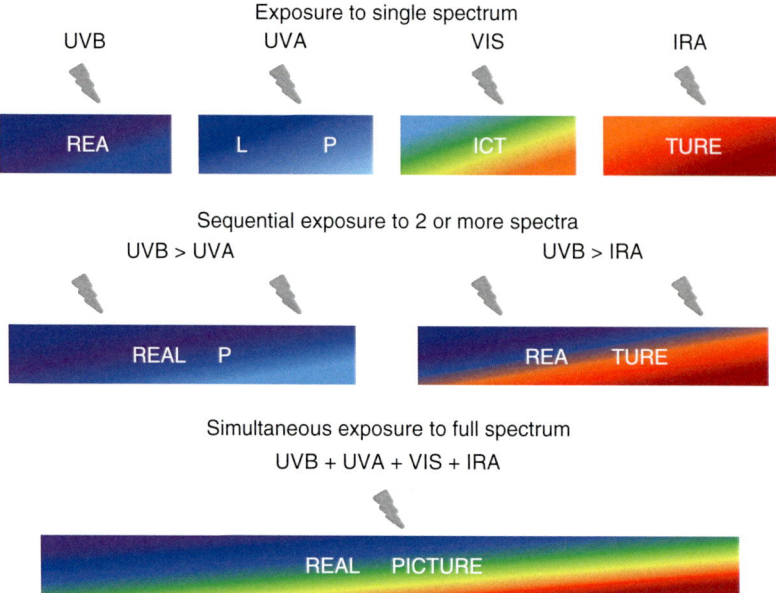

Fig. 2.1 Impact of different exposure protocols on the quality of the picture of cutaneous responses to solar light. Exposure of the skin (or any form of cutaneous test system) to either UVB, UVA, VIS, or IRA radiation alone will produce four different pictures each containing only limited information. Sequential irradiation with two or more sub-spectra (e.g., UVB first, then IRA) will generate another set of pictures that may contain/lack certain aspects of the respective single exposure responses but also include new information. However, the "real picture" displaying the whole multitude of adaptive and maladaptive responses of the skin toward solar radiation will only be drawn in response to a simultaneous exposure to the full spectrum of sunlight, i.e., UVB plus UVA plus VIS plus IRA, when applied in (patho)physiological relevant fluencies and ratios

in order to fully understand the biological impact of solar radiation on human skin (Fig. 2.1).

2.2 Effects of Near-Infrared Radiation (IRA) on Human Skin

It is now generally accepted that similar to UVB or UVA radiation, IRA radiation may exert profound biological effects at human skin in general and the dermal compartment of the skin, in particular (reviewed in [1, 3]). As it is the case with UV radiation, IRA-induced effects may be harmful, e.g., by contributing to photoaging but, under certain conditions, also beneficial, e.g., when used therapeutically to treat sclerotic skin lesions or to stimulate wound healing. As we are here primarily concerned about IRA-induced detrimental effects, we will not discuss the potential therapeutic use of this type of radiation.

The impact of IRA radiation on human skin is best illustrated by the recent observation that this type of radiation changes the transcriptome of primary human skin fibroblasts [4]. In this study, approximately 600 genes were found to be IRA responsive, and functional clustering of these genes came under groups involved in extracellular matrix homeostasis, apoptosis, cell growth, and stress responses [4]. These genes are in a broader sense related to photoaging and possibly also photocarcinogenesis. Accordingly, among the genes that were significantly upregulated in primary human skin fibroblasts was Matrix metalloproteinase-1 (MMP-1), thus confirming a previous report of IRA-induced MMP-1 mRNA expression in this cell type [5]. Increased MMP-1 mRNA expression occurred in human skin fibroblasts without a concomitant upregulation of its tissue specific inhibitor TIMP-1, indicating the possibility that IRA radiation may cause increased MMP-1 activity and thus breakdown of collagen fibers, which would ultimately cause the formation of coarse wrinkles and thus a clinical hallmark of photoaged skin [5]. In fact, it is now generally accepted that IRA radiation is causally related to wrinkle formation in the skin because the original in vitro observation of IRA-induced MMP-1 upregulation was shown to be of in vivo relevance both for human [6] and mouse [7] skin. Even more important, chronic exposure of hairless mice to IRA radiation caused the formation of coarse wrinkles [7], and the combined treatment of these animals with IRA plus UV radiation resulted in wrinkle formation which was more than that achieved by UV alone or IR alone, indicating that the two types of radiation were causing photoaging through different (photobiological, molecular) mechanisms. Increased MMP-1 expression is presumably not the only mechanism responsible for IRA-induced photoaging because IRA exposure was also reported to reduce type I collagen expression by reducing the production of procollagen-1-stimulating transforming growth factor-ß1, transforming growth factor-ß2, and transforming growth factor-ß3 expression in human skin [8]. Also, IRA radiation may induce angiogenesis in human skin by a mechanism involving the increased expression of vascular endothelial growth factor [9], i.e., another molecular feature of photoaged skin [25], which was suggested to be of functional relevance for photoaging-associated wrinkle formation [10]. In addition, IRA radiation was found to increase the number of mast cells in human skin in vivo [11], and this effect is also indicative of its potential to cause photoaging. As previously mentioned [3], it should be noted that the majority of these studies have employed artificial irradiation devices, which do not mimic natural sunlight but emit preferentially or even selectively IRA radiation. It has therefore been argued that the findings described above are of limited relevance for natural IRA irradiation during sun exposure [12]. In fact, in some studies employing artificial irradiation devices, the IRA radiation doses used could also be criticized for exceeding the physiological doses the skin usually receives when an individual is exposed to the sun. Nevertheless, similar MMP-1 production has been reported in more recent studies using low and repetitive IR doses. Thus, regular exposure to IRA

could be more important than expected in premature skin aging [13]. The physiological relevance of IRA-induced skin aging in humans is also strongly emphasized by a very elegant study by Cho et al. [14], in which the effects of natural sunlight, sunlight minus UVR, or the heat component within natural sunlight only were directly compared with each other. By exposing human buttock skin to the three different qualities of natural sunlight, it was shown that UV-filtered sunlight significantly increased MMP-1 expression in exposed skin, indicating that IRA radiation contributes to natural sunlight-induced skin responses. Collectively, these independent studies provide clear evidence that IRA radiation causes photoaging.

In contrast to photoaging, the role of IRA radiation in photocarcinogenesis is less well studied. Accordingly, IRA irradiation if provided prior to UVB radiation may prevent UVB-induced keratinocyte apoptosis (sunburn cell formation) and might thereby contribute to the development of skin cancer [15]. This assumption is further supported by a recent in vivo photocarcinogenesis study from the same authors, in which IRA irradiation preceding UVB irradiation did not cause earlier or more tumor growth but was associated with significantly accelerated, more aggressive tumor growth and a higher number of more malignant skin tumors [16]. Further studies are clearly required to answer the question if exposure of the skin to IRA radiation is associated with an increased risk of developing nonmelanoma and/or melanoma skin cancer.

2.3 Effects of Visible Light on Human Skin

Visible light wavelengths range from violet (400 nm) to profound red (740 nm). In contrast to the numerous studies which were conducted to analyze the effects of IRA radiation on the skin, the number of studies focusing on the visible light and skin is still limited to a very few. Accordingly, Zastrow et al. by means of electron spin resonance demonstrated increased free radical formation in ex vivo irradiated human skin as a consequence of exposure to wavelengths not only in the UV and IRA but also in the visible range [17, 18]. These observations have recently been extended in a study in which direct free radical production was measured using EPR spectrophotometry ex vivo and in vivo [19]. It was found that, over the time, radical species cumulate, if the skin was exposed to UV (325–380 nm), visible light, and IRA irradiation. Importantly, radical species production was significantly stronger in vivo compared to ex vivo. This study also revealed that some stratum corneum lipids are modulated after this exposure. After UVR, the ceramide subclass [AP2] decreased and the ceramide subclass [NP2], sodium cholesteryl sulfate (SCS), and squalene (SQ) increased. Conversely, after VL and IR irradiations, ceramide [AP2] and SCS increased and SQ significantly decreased.

Biological consequences of visible light irradiation of human skin were shown by Mahmoud et al. [20]. By employing an artificial irradiation device with an emission mainly confined to wavelengths between 400 and 800 nm, these authors were first to provide clear-cut evidence that wavelengths in the visible range at doses between 80 and 480 m Watts/cm^2 can cause pigmentation in vivo in human skin in the absence of UV radiation. Visible light-induced skin pigmentation was apparent at the clinical level and confirmed by histopathology. Interestingly, visible light-induced skin pigmentation was only found to occur in darker skin types, i.e., skin types >4, according to Fitzpatrick. These observations were recently confirmed and extended in independent studies [21], in which the propigmenting properties of blue-violet light (415 nm) were compared to red light (630 nm) on dorsal skin of individuals with type III and IV skin types. It was found that the blue-violet light induced a marked and prolonged dose-related pigmentation at physiological doses, whereas the red light did not induce any pigmentation. The mechanisms responsible for blue-violet light-induced skin pigmentation are currently unknown but appear to be different from the ones involved in UVB-induced skin pigmentation. There is also indirect evidence that exposure to visible light can worsen melasma. In a clinical study, a sunscreen providing protection against UVB/UVA plus visible light proved to be superior to a control sunscreen with identical UVB/UVA protection but without protection against visible light in the prevention of melasma [22].

There is also evidence that visible light exposure might contribute to photoaging by enhancing collagen breakdown [23]. In this study, in vitro exposure of human epidermis models to visible light was reported to increase MMP-1 as well as TNF-alpha mRNA expression in epidermal keratinocytes. This gene regulatory activity of visible light was associated with an increased production of reactive oxygen species (ROS) in these epidermis models, and this latter observation could be confirmed in vivo in human skin when ROS production was measured by Raman spectroscopy. Experimental evidence from human in vivo studies that visible light possesses gene regulatory activities and from in vivo animal studies that chronic exposure to visible light indeed causes wrinkle formation in the skin is currently lacking. In addition, virtually nothing is known about the role of visible light in skin carcinogenesis. Nevertheless, the existing studies are in line with the assumption that visible light might exert some biological effects on human skin which include an increase in ROS production and which are of relevance for skin pigmentation (and maybe photoaging) at least in darker pigmented individuals.

2.4 Different Wavelengths Within and Beyond the UV Spectrum Interact with Each Other

The vast majority of published studies has exclusively analyzed each wavelength range, i.e., UVB or UVA or visible light or IRA radiation-induced effects on human skin, separately. Given that human skin is naturally exposed to all of these wavelengths simultaneously, because they are all part of natural sunlight, it is conceivable to assume that interactions between the different responses elicited by each wavelength range may exist. Support of this concept was first provided by Schieke et al. [24], who demonstrated a molecular cross talk between

UVA and UVB signaling in human epidermal keratinocytes at the level of MAPK activation. Accordingly, UVA radiation alone was found to cause a modest and transient activation of ERK1/2 15–30 min after exposure, whereas UVB irradiation caused a strong and immediate ERK1/2 phosphorylation that lasted for up to 1 h. Only minor activation of p38 and JNK1/2 was detected after both UVA and UVB irradiation. A different pattern was observed, if keratinocytes were sequentially exposed, i.e., first to UVA followed immediately by UVB exposure. In this case, the UVB-induced strong phosphorylation of ERK1/2 was prevented, but instead p38 and JNK phosphorylation were enhanced. Of note, this activation pattern was also observed, if the sequence was altered, i.e., if keratinocytes were first irradiated with UVB and then immediately thereafter with UVA. In aggregate, these results strongly indicate that UVA and UVB irradiation cause distinct stress responses in keratinocytes and that sequential elicitation of these two stress responses causes a third response that is different from either alone and cannot be explained by a simple addition of effects. The authors therefore concluded that the molecular cross talk of UVA and UVB irradiation which they had observed at the level of MAPK signaling represents an evolutionary conserved signaling pathway, which may have developed as an elaborate molecular defense strategy of human skin cells to respond to solar radiation-induced stress in a way which goes beyond a mere additive effect of its single components, i.e., in this case UVA and UVB. Indeed, there is more evidence in the literature that cross talk signaling may also occur for UVB and IRA radiation, although in this case the response even differs if the sequence of irradiations is being changed from first IRA, then UVB to first UVB, then IRA (reviewed in [1]).

2.5 A Novel Approach to Better Understand the Effects of Natural Sunlight on Human Skin

These examples emphasize the need for more detailed analysis of the relative contribution of each wavelength to the net biological effect, which is caused by natural sunlight in human skin cells.

We believe that this challenge can be best met by the development of a novel irradiation device, which (1) allows simultaneous exposure of human skin cells and the skin to UVB, UVA, visible light, and IRA at physiologically relevant dose levels and thus the study of the "natural UV stress response" but which (2) also permits to dim off selected wavelength areas to better understand their relative contribution to that response. Studies employing such an irradiation device will be of enormous clinical relevance for efficient photoprotection of human skin. If one accepts that during evolution human skin has adapted to natural sunlight and thus the exposure to a combination of different wavelengths within and beyond the UV spectrum, with the overall goal to provide an optimized stress response, which serves to limit skin damage as much as possible, then the study of stress responses elicited in human skin by exposure to single wavelength areas only or by irradiation protocols which merely sequentially add two or more types of irradiation may lead to results which are (1) of no or only limited physiological relevance (2) and thereby misleading when it comes to the development of photoprotective measures for human skin. As a consequence, current sunscreen products may not yet be optimal.

Acknowledgments This work was supported by BMBF (Förderkennzeichen 02NUK036CKAUVIR TP C JK).

References

1. Grether-Beck S, Marini A, Jaenicke T, Krutmann J. Photoprotection of human skin beyond ultraviolet radiation. Photodermatol Photoimmunol Photomed. 2014;30:167–74.
2. Krutmann J, Bouloc A, Sore G, Bernard BA, Passeron T. The skin aging exposome. J Dermatol Sci. 2016;85(3):152–61. doi:10.1016/j.jdermsci.2016.09.015.
3. Krutmann J, Morita A, Chung JH. Sun exposure: what molecular photodermatology tells us about its good and bad sides. J Invest Dermatol. 2012;132 (3 Pt 2):976–84.
4. Calles C, Schneider M, Macaluso F, Benesova T, Krutmann J, Schroeder P. Infrared a radiation influences the skin fibroblast transcriptome:

mechanisms and consequences. J Invest Dermatol. 2010;130:1524–36.

5. Schieke S, Stege H, Kurten V, Grether-Beck S, Sies H, Krutmann J. Infrared-a radiation-induced matrix metallo-proteinase 1 expression is mediated through extracellular signal regulated kinase 1/2 activation in human dermal fibroblasts. J Invest Dermatol. 2002;119:1323–9.

6. Schroeder P, Lademann J, Darvin ME, et al. Infrared radiation-induced matrix metalloproteinase in human skin: implications for protection. J Invest Dermatol. 2008;128:2491–7.

7. Kim HH, Lee MJ, Lee SR, et al. Augmentation of UV-induced skin wrinkling by infrared irradiation in hairless mice. Mech Ageing Dev. 2005;126:1170–7.

8. Kim MS, Kim YK, Cho KH, Chung JH. Regulation of type I procollagen and MMP-1 expression after single or repeated exposure to infrared radiation in human skin. Mech Ageing Dev. 2006;127:875–82.

9. Kim MS, Kim YK, Cho KH, Chung JH. Infrared exposure induces an angiogenic switch in human skin that is partially mediated by heat. Br J Dermatol. 2006;155:1131–8.

10. Detmar M. The role of VEGF and thrombospon-dins in skin angiogenesis. J Dermatol Sci. 2000;24 (Suppl. 1):S78–84.

11. Kim MS, Kim YK, Lee DH, et al. Acute exposure of human skin to ultraviolet or infrared radiation or heat stimuli increases mast cell numbers and trypt-ase expression in human skin in vivo. Br J Dermatol. 2009;160:393–402.

12. Piazena H, Kelleher DK. Effects of infrared-a irradia-tion on skin: discrepancies in published data highlight the need for an exact consideration of physical and photobiological laws and appropriate experimental settings. Photochem Photobiol. 2010;86:687–705.

13. Robert C, Bonnet M, Marques S, Numa M, Doucet O. Low to moderate doses of infrared a irradiation impair extracellular matrix homeostasis of the skin and contribute to skin photodamage. Skin Pharmacol Physiol. 2015;28:196–204.

14. Cho S, Lee MJ, Kim MS, et al. Infrared plus visible light and heat from natural sunlight participate in the expression of MMPs and type I procollagen as well as infiltration of inflammatory cell in human skin in vivo. J Dermatol Sci. 2008;50:123–33.

15. Jantschitsch C, Majewski S, Maeda A, Schwarz T, Schwarz A. Infrared radiation confers resistance to UV-induced apoptosis via reduction of DNA damage and upregulation of antiapoptotic proteins. J Invest Dermatol. 2009;129:1271–9.

16. Jantschitsch C, Weichenthal M, Maeda A, Proksch E, Schwarz T, Schwarz A. Infrared radiation does not enhance the frequency of ultraviolet radiation-induced skin tumors, but their growth behaviour in mice. Exp Dermatol. 2011;20:346–50.

17. Zastrow L, Groth N, Klein F, et al. The missing link – light-induced (280–1600 nm) free radical formation in human skin. Skin Pharmacol Physiol. 2009;22:31–44.

18. Zastrow L, Groth N, Klein F, Kockott D, Lademann J, Ferrero L. UV, visible and infrared light: which wavelengths produce oxidative stress in human skin? Hautarzt. 2009;60:310–7.

19. Lohan SB, Muller R, Albrecht S, et al. Free radi-cals induced by sunlight in different spectral regions—in vivo versus ex vivo study. Exp Dermatol. 2016;25:380–5.

20. Mahmoud BH, Ruvolo E, Hexsel CL, et al. Impact of long-wavelength UVA and visible light on melano-competent skin. J Invest Dermatol. 2010;130:2092–7.

21. Duteil L, Cardot-Leccia N, Queille-Roussel C, et al. Differences in visible light-induced pigmentation according to wavelengths: a clinical and histological study in comparison with UVB exposure. Pigment Cell Melanoma Res. 2014;27(5):822–6.

22. Boukari F, Jourdan E, Fontas E, Montaudié H, Castela E, Lacour JP, Passeron T. Prevention of melasma relapses with sunscreen combining protection against UV and short wavelengths of visible light: a pro-spective randomized comparative trial. J Am Acad Dermatol. 2015;72(1):189–190.e1.

23. Liebel F, Kaur S, Ruvolo E, Kollias N, Southall MD. Irradiation of skin with visible light induces reac-tive oxygen species and matrix-degrading enzymes. J Invest Dermatol. 2012;132:1901–7.

24. Schieke SM, Ruwiedel K, Gers-Barlag H, Grether-Beck S, Krutmann J. Molecular crosstalk of the ultra-violet A and ultraviolet B signaling responses at the level of mitogen-activated protein kinases. J Invest Dermatol. 2005;124(4):857–9.

25. Yano K, Oura H, Detmar M. Targeted overexpression of the angiogenesis inhibitor thrombospondin-1 in the epidermis of transgenic mice prevents ultraviolet-B induced angiogenesis and cutaneous photo-damage. J Invest Dermatol. 2002;118:800–5.

The Impact of Climate Change on Skin and Skin-Related Disease

3

Louise K. Andersen

3.1 Climate Change and Its Consequences

Global warming refers to a rise in average global temperature at the Earth's surface. The primary cause of global warming is anthropogenic emissions of greenhouse gases into the atmosphere. The majority of these gases come from the burning of fossil fuels to produce energy. Deforestation and industrial/agricultural processes also emit greenhouse gases. The average surface temperature has increased by 0.6 °C over the past 100 years [1], and another 2 °C rise in temperature is expected by the end of this century [2]. Global warming causes changes in weather patterns and climate and represents one aspect of climate change. Climate change refers to changes in the statistical distribution of weather patterns for an extended period of time, and it has been associated with a rise in temperature, an increase in the frequency of extreme weather events, (e.g., heavy rainfall, floods, droughts, and hurricanes), and a rise in sea levels. Changes in ecology in response to climate change lead to a shift in the geographical distribution and behavior of insects and pathogens [3].

Climate change will affect human health, mostly adversely, and cause disease outbreaks. The information on how climate change will impact on skin and skin-related diseases is growing [4–7]. The direct impacts of climate change on the skin include the effects of extreme weather events such as heavy rainfall, floods, droughts, and hurricanes, which can lead to an increase in skin infections, inflammatory skin diseases, and traumatic skin disorders. The indirect effects of climate change arise from the disruption of natural systems leading to an increase in the incidence of vector-borne and waterborne diseases with many of them causing skin manifestations.

3.2 Skin Infections, Inflammatory Skin Diseases, and Traumatic Skin Disorders

With climate change, extreme weather events such as heavy rainfall, floods, droughts, and hurricanes are predicted to occur with increasing frequency and greater intensity. Flooding is often associated with hurricanes, tsunamis, or heavy precipitation and is the most common natural disaster in both developed and developing countries. Floods may have devastating consequences on buildings, roads, and the transport system. The unhygienic conditions, crowded living conditions, exposure to water or friction, and contact with objects during and after flooding may result

L.K. Andersen, M.D.
Department of Dermato-Venereology, Aarhus
University Hospital, Aarhus C, Denmark
e-mail: loand02@privat.dk

© Springer International Publishing Switzerland 2018
J. Krutmann, H.F. Merk (eds.), *Environment and Skin*,
https://doi.org/10.1007/978-3-319-43102-4_3

in an upsurge of disease among affected [8, 9]. Skin disease occurs in victims, workers, and others who come into contact with the contaminated floodwater after the disaster [10, 11]. Floodwater may contain infectious organisms including bacteria and viruses, dangerous materials such as industrial chemicals, as well as sharp objects (e.g., glass or metal fragments) that can cause skin injuries. Buildings that have been damaged by floodwater can pose health risks from molds, insects, and chemicals.

Injured skin exposed to floodwaters can become infected by staphylococci (including methicillin-resistant *Staphylococcus aureus* or MRSA), *Streptococcus aureus*, *Aeromonas* spp., as well as other uncommon pathogens such as pseudomonas spp., *Burkholderia pseudomallei* (melioidosis), and rapid-growing mycobacteria [12–14]. Tetanus can be acquired from contaminated soil or water entering breaks in the skin, such as abrasions and cuts. Infections with marine *Vibrio* species (*V. vulnificus* and *parahaemolyticus*) can result if injuries are contaminated with ocean water [15]. The prevalence of traumatic skin disorders is highest in tsunami survivors [9]. Prolonged exposure to floodwater can increase the risk of fungal skin infections. Floodwater can contribute to keratinocytes damage conducting to inflammation and irritation of the skin, while increasing the risk of eczema, particularly irritant contact dermatitis. Other inflammatory skin diseases reported after flooding have included xerosis, prurigo, chronic urticaria, and seborrheic dermatitis among others [8]. After the flooding in Thailand in 2006, the most common skin problems were skin infections followed by inflammatory skin diseases and traumatic skin disorders. In addition, the psychological stress of natural disasters can trigger or cause aggravation of an underlying skin disease such as psoriasis, dermatitis, urticaria, alopecia areata, or vitiligo [16, 17].

3.3   Waterborne Disease

Waterborne diseases are bacterial, viral, or parasitic diseases, which are transmitted through contaminated freshwater. Infection can occur after ingestion of contaminated water directly or through food or via skin contact with the water (e.g., bathing, washing). Waterborne diseases are one of the major causes of morbidity and mortality, particularly in developing countries [18].

Heavy rainfall and flooding can spread bacteria, sewage, and other organic waste into waterways and aquifers, leading to disease outbreaks. Higher temperatures can also exacerbate water contamination, as rising temperatures encourage the growth bacteria and other microorganisms. Heavy rainfall and flooding can increase the risk of waterborne disease, with certain of them causing skin manifestations.

3.3.1   Leptospirosis

Leptospirosis, transmitted by rodents through their urine, is commonly reported in the tropical areas. People are infected through contact with contaminated freshwater or mud. Infected people can develop a petechial rash and later (in the second phase), in severe cases, jaundice [19]. Urban epidemics of leptospirosis often occur in unsanitary environments after heavy rain and flooding. For example, floods in Thailand in 2011 caused outbreaks of leptospirosis in the Bangkok Metropolitan Region [20]. In the District of Anuradhapura, Sri Lanka, a sevenfold increase in rainfall with subsequent flooding caused leptospirosis outbreaks [21]. Throughout December 2010 and January 2011, Queensland, Australia, experienced widespread flooding as a consequence of unusually heavy rainfall resulting in acute cases of leptospirosis in people exposed to the contaminated floodwater [22]. After a typhoon with subsequently flooding in September 2009, an outbreak of leptospirosis occurred in the Philippines [23]. Unusually high rainfall followed by flooding also caused outbreaks of leptospirosis in Guyana, South America 2005 [24].

3.3.2   Schistosomiasis

Schistosomiasis is spread by contact with water that carries the parasitic worms of the Schistosoma type. Freshwater snails that have been infected release these parasites. Schistosomiasis occurs in Asia, Southern America, as well as Africa. Infected persons often develop cercarial dermatitis, particularly on the feet, after contact with contaminated water [25]. In China, along the Yangtze River, the number of acute cases of schistosomiasis is reported to be markedly higher in years with flooding (on average, 2.8 times more cases) compared to the years with normal water levels [26]. Heavy rainfall and flooding in Porto de Galinhas, Pernambuco, Brazil, caused outbreaks of schistosomiasis in 2000 as well as in the following years [27]. A higher temperature associated with climate change is reported to affect the geographical distribution of certain types of schistosoma. In China, for example, it has been predicted that schistosomiasis can reemerge in areas where transmission has been successfully interrupted and in the future expand into currently non-endemic areas in the northern part of the country [28].

3.4   Vector-Borne Diseases

The incidence of vector-borne disease is increasing globally. Vector-borne diseases are infections spread by the bites of infected arthropods. The arthropods that most typically serve as vectors are mosquitoes, ticks, triatomine bugs, and sand flies. Vectors become infected by a disease agent as feeding on infected vertebrates (e.g., birds, deers, and other animals) and then transmitted the microbe to a susceptible person.

High temperatures affect disease vectors by influencing their population density and survival rate, changing their susceptibility to pathogens and their geographical distribution (rising to higher altitudes) [29–33]. More breeding sites and increased sizes of vector population (e.g., mosquitoes) are reported after increased rainfall. Climate change can lead to outbreaks of certain vector-borne disease, with many of them causing skin manifestations.

3.4.1   Dengue

Dengue is one of the most widespread diseases and causes over 50 million infections annually worldwide. Dengue is transmitted by the bite of mosquito (*Aedes aegypti* and *Aedes albopictus*). Dengue occurs in Africa, Southeast Asia, and China as well as countries in the Pacific Ocean and America (Fig. 3.1) [34]. Those infected with dengue typically develop a centrifugal

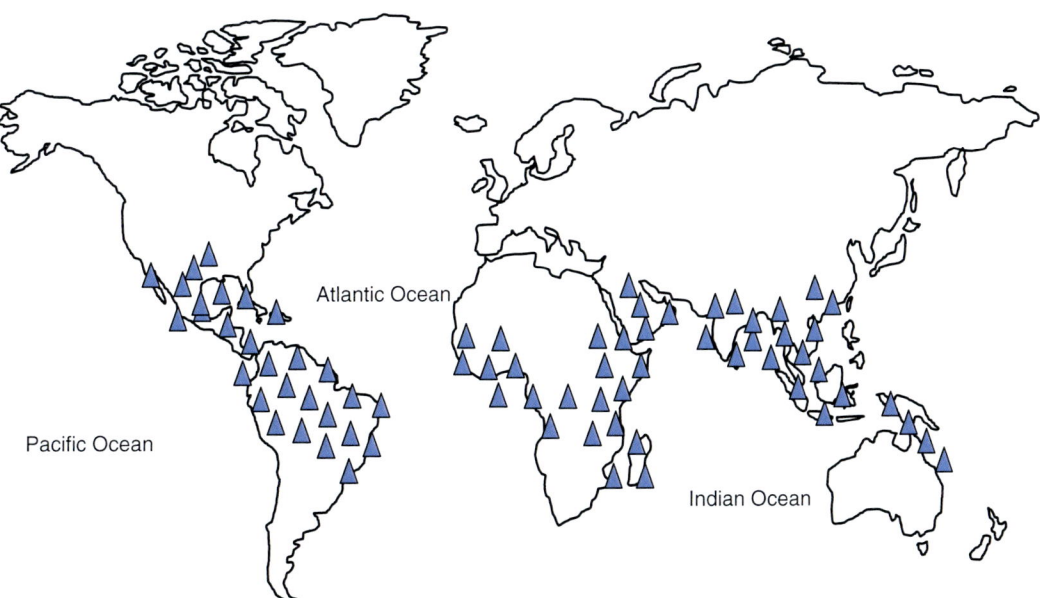

Fig. 3.1 Risk areas for dengue transmission. *Triangles* indicate these areas

maculopapular rash. Others may develop a pete-chial or scarlatiniform rash starting on the dor-sum of the hands and then spreading to the arms, the legs, as well as the torso [35].

A higher temperature as a consequence of cli-mate change can lead to a shift in the geographi-cal distribution of dengue. In Europe, for example, dengue can spread to unaffected areas such as the coastal areas of the Mediterranean and Adriatic seas and the Northeastern part of Italy [36]. Other unfavorable areas, such as inland Australia, the Arabian Peninsula, southern Iran, and some parts of North America, can become climatically favorable for *A. aegypti* in the future. Projections based on future climate scenarios could potentially see a reduction in distribution in other parts of the world [29]. In Guangzhou, China, the number of dengue cases was posi-tively associated with mosquito density, imported cases, temperature, precipitation, vapor pressure, and minimum relative humidity, while being neg-atively associated with air pressure [37]. In Dhaka, Bangladesh, the dengue incidence was predicted to increase more than 40 times by the year 2100 compared to 2010, if the ambient tem-peratures rose by 3.3 °C, according to the IPCC regional climate projection. The maximum tem-perature and relative humidity were major deter-minants of dengue transmission in Dhaka for the period of 2000–2010 [38]. In Singapore, the den-gue incidence increased linearly 5–16 and 5–20 weeks following elevated temperature and pre-cipitation, respectively [39]. In the city of Cairns, Queensland, in tropical Australia it seems unclear whether dengue incidence will increase under climate change scenarios (2046–2064) based only on vector factor [40].

3.4.2 West Nile Virus

West Nile virus (WNV) belongs to the genus *Flavivirus* and is transmitted by mosquitoes. Infected people occasionally develop a maculo-papular rash [41]. WNV infections were first reported in Uganda in 1937. Afterward, the WNV spread across Africa to the Middle East, Asia, and Eastern Europe [42]. The first case of human WNV infection in the United States was reported in 1999. Since then, WNV infection has spread across North America and has become a major public health concern [43].

The increasing temperatures during the com-ing decades and inevitable changes in the fauna and flora may affect the geographical distribution of WNV. For example, in North America it is pre-dicted that WNV may be transmitted at more northern altitudes in the future [44, 45]. A 5 °C increase in mean maximum weekly temperature has been associated with a 32–50% higher inci-dence of WNV infection in the United States. The presence of at least 1 day of heavy rainfall within a week has been associated with a 29–66% higher incidence of WNV infection during the same week or over the following 2 weeks [46]. Outbreaks of human WNV infections have also been associated with droughts, which may increase in frequency as a consequence of global climate change [43, 47].

3.4.3 Chikungunya

Chikungunya (CKG), caused by the chikungunya virus, is transmitted by mosquitoes of the genus *Aedes* (*A. aegypti* and *A. albopictus*). CKG cases have been reported in Africa, Southeast Asia, America, and certain part of Europe (Fig. 3.2) [48]. There are many cutaneous manifestations in those infected with CKG. The most common is an erythematous maculopapular rash on the trunk, limbs, and face. Small vesicles and bullae may occur [49].

An increase in temperature above the average temperature appears to increase the proportion of virus-susceptible mosquitoes in the population. On the other hand, if fresh larva are exposed to temperatures greater than 44.5 °C for 10 min, the mortality rate is found to be approximately 95%. Temperatures greater than 45 °C for 10 min or more are lethal for larva [30].

Rain is found to be a dominant factor for the spatiotemporal transmission of CKG [50]. During an epidemic in central Thailand, the num-ber of CKG cases rose 6 weeks after increasing cumulative rainfall, with variation of average

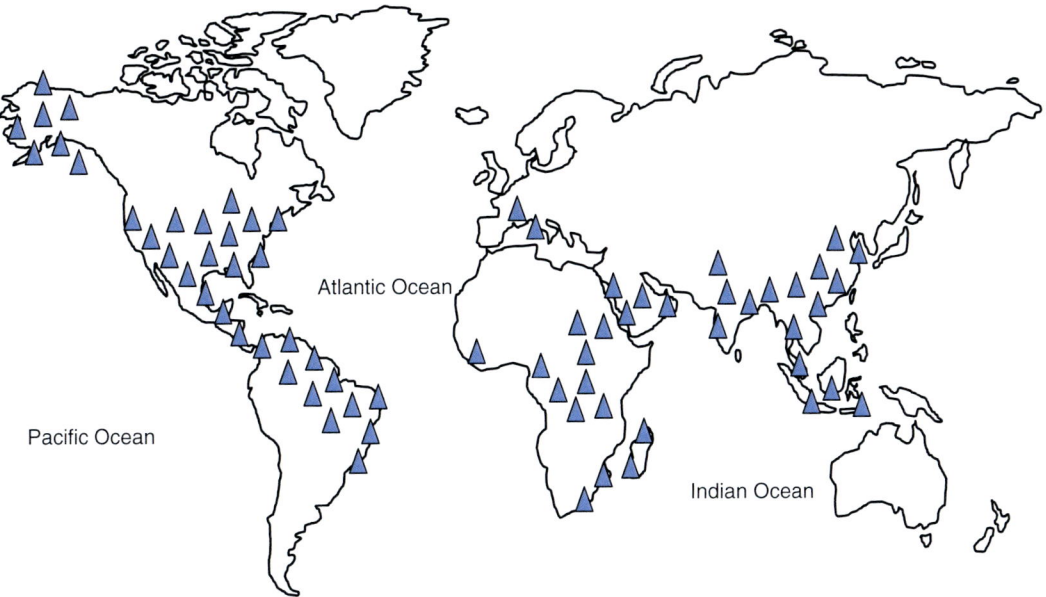

Fig. 3.2 Risk areas for chikungunya transmission. *Triangles* indicate these areas

daily temperatures (23.7–30.7 degrees) per week. Acute cases of CKG also spread from south to north Thailand with a median speed of 7.5 km per week [51]. Outbreaks of CKG along coastal Kenya have been associated with unusually dry conditions [52].

3.4.4 Ross River Virus

The *Ross River virus* (RRV), which is transmitted by mosquitoes, occurs in Australia. A maculo-papular rash may occur in those infected. Purpuric lesions and small vesicles may also occur [53].

Rainfall affects the transmission of RRV, as mosquitoes require water to support the larval and pupal stages of development. A higher minimum temperature in the winter also assists larvae survival. Outbreaks of RRV disease in Australia have been associated with temperature, rainfall, relative humidity, as well as the Southern Oscillation Index [54, 55]. Epidemics of RRV are typically observed in temperate areas of Australia with heavy rainfall and flooding. In the tropical areas, RRV transmission occurs throughout the year. Weather forecasts can be used in conjunction with other surveillance techniques, to identify conditions suitable for an epidemic of RRV with a high degree of accuracy [56].

3.4.5 Chagas Disease

Chagas disease (or American trypanosomiasis) a tropical parasitic disease caused by the protozoan *Trypanosoma cruzi*. *Trypanosoma cruzi* is transmitted by the bite of bugs of the subfamily Triatominae. The diseases can also be transmitted after ingestion of contaminated food. Cutaneous manifestations in those infected depend on the transmission route and phase of the disease. Cutaneous manifestations may be indurated erythematous plaques with necrosis, erythematous papules and nodules, panniculitis or skin ulcerations, chagoma, Romaña sign, and diffuse or morbilliform rashes [35].

The increasing temperatures during the coming decades and changes in ecological conditions can lead to a shift in the geographical distribution of Chagas disease. For example, in the Southern Texas and North Mexico, the most common vectors of *Trypanosoma cruzi* (*T. gerstaeckeri* and

T. sanguisuga) are predicted to shift their geographical distribution to higher altitudes in the future [57]. Under ecological niche models, the *Triatoma brasiliensis* have shown to have the potential to colonize other geographical areas of Brazil [58]. Increasing temperatures are also predicted to affect the geographical distribution of Chagas disease in Colombia [59]. Hurricane Isidore, which struck the Yucatán Peninsula of Mexico, increased the abundance of *T. dimidiata* between 10- and 15-fold and caused disease outbreaks for about 6 months after the disaster [60].

3.4.6 Tularemia

Tularemia, endemic in North America and parts of Europe and Asia, is caused by the bacterium *Francisella tularensis*. Its primary vectors are ticks and deer flies. The disease can, however, also be spread through other arthropods or by inoculation through the skin, oropharyngeal mucosa, conjunctiva via contaminated water, blood, or tissue. In cases with skin symptoms, a small, erythematous, tender, or pruritic papule will develop at the site of inoculation; 2–3 days after the papule enlarges and ulcerates [61].

With the increasing temperatures during the coming decades, inevitable changes in the fauna and flora may affect the geographical distribution of tularemia. For example, in the United States, it is predicted that tularemia may be transmitted at more northern altitudes in the future [62]. A high temperature followed by rain caused outbreaks of tularemia in Kayseri, Turkey, in 2010 and 2012 [63]. In Sweden, an approximately 2 °C increase in the monthly average summer temperatures was predicted to cause outbreaks of tularemia in the following weeks, ranging from 3.5 to 6.6 weeks, depending on the geographical area [64].

3.4.7 Lyme Disease

Lyme disease, is transmitted to humans by the bites of ticks. The tick, *Ixodes ricinus*, is the primary European vector, whereas the *Ixodes scapularis* is the primary vector in North America. In those infected, the erythema migrans rash can occur, 1 day–1 month after being bitten [65].

Over the last few decades, the incidence of Lyme disease has increased in parts of Europe as well as North America. The increasing temperatures are creating hospitable climate conditions for the survival and expansion of vectors. In Scandinavia (Sweden, Norway, and Finland), northward extension of the range and abundance of *I. ricinus* is projected for the periods 2011–2040, 2041–2070, and 2071–2100 [66]. In North America, the potential effects of climate change on *I. scapularis* have shown that the vector may retract from southern United States and spread into the central United States as well as into south-central Canada [67–69]. In Hungary, the incidence of Lyme disease did increase significantly between 1998 and 2010 and was shown to be associated with the weekly temperature from weeks 1 to 23 [70].

3.4.8 Leishmaniasis

Leishmaniasis, caused by protozoan parasites of the genus *Leishmania*, is transmitted to human by sand flies. Cutaneous leishmaniasis occurs in Southern America, Africa, Asia, and Southern Europe (Fig. 3.3) [71]. Cutaneous leishmaniasis is the most common type and presents with a red furuncle or papule at the bite site, which progresses to a skin ulcer within months [35].

The number of leishmaniasis cases is increasing worldwide, especially in North America and Europe. Altitudinal patterns of sand flies and host abundance are predicted to change as a consequence of ongoing climate change. Based on climate scenarios for 2020, 2050, and 2080 in North America (Mexico, the United States, and Canada), a northward range expansion is expected from Mexico and the southern United States and possibly even in parts of south-central Canada, as more habitat becomes suitable for sand flies and reservoir species [72]. Projections of potential geographic distributions across climate scenarios suggest that *Lutzomyia whitmani*, one of the vectors of cutaneous leishmaniasis in Southern Brazil, could have the potential to

Fig. 3.3 Risk areas for cutaneous leishmaniasis transmission. *Triangles* indicate these areas

extend its geographic range in Southeastern Brazil, where leishmaniasis appears to reemerge [73]. In the pre-Saharan zones of North Africa, increased temperatures and precipitation could result in an increased risk of cutaneous leishmaniasis. Temperature above a critical range appears, however, to decrease the risk of cutaneous leishmaniasis, apparently by limiting the reproductive activity of the disease vectors [74]. In Europe, changes in climate contribute to a shift in the geographical distribution of leishmaniasis. Under climate change scenarios, the Central European climate is predicted to become more suitable for sand flies with a current southwestern focus [75]. Under simulated climate change in the Madrid Region in Spain, a three-fold increase in apparent vector densities was predicted from 2070 to 2100 compared with the current situation [76].

3.5 Preventions and the Future

Climate change has become one of the most pressing global issues in the twenty-first century. It can increase the risk of certain skin and skin-related diseases, particularly in developing countries, which have poor healthcare assessment.

After floods disease may be prevented and controlled by providing a safe water supply, implementing good sanitation, and the provision of field hospitals and shelters. Safe food preparation, nutrition, and medical supplies as well as access to fuel and transportation are needed. Education of the public and healthcare workers about disease risk and prevention could also benefit [77].

Vector control is essential. For example, the ability to predict outbreak of a certain vector-borne disease, months or even years in advance based upon climatic indicators, makes early intervention initiatives possible. Geographic information systems designed to capture all types of spatial or geographical data can help to address the behavior of disease vectors in the future. The most effective way to prevent disease caused by the bite of vectors could, however, be indoor residual spraying, mosquito repellent, insecticide-treated nets, as well as traps [77].

In summary, there is increasing evidence that climate change affects the occurrence of skin and skin-related diseases, particularly in developing countries. A better awareness of the effects of climate change could help to reduce outbreaks of skin and skin-related disease; however, the ultimate goal for the future is to reduce the emissions of greenhouse gases to prevent further climate change, if possible.

References

1. Houghton JT, Ding Y, Griggs DJ, et al., editors. Climate change 2001: the scientific basis: contribution of working group i to the third assessment report of the intergovernmental panel on climate change. Cambridge: Cambridge University Press; 2001.
2. Houghton JT, Meira Filho LG, Callander BA, et al., editors. Climate change 1995: the science of climate change: contribution of working group i to the second assessment report of the intergovernmental panel on climate change. Cambridge: Cambridge University Press; 1996.
3. United States Enviromental Protection Agency. http://www.Epa.Gov/climatechange/impacts-adaptation/health.Html. Accessed 3 Jan 2015.
4. Andersen LK. Global climate change and its dermatological diseases. Int J Dermatol. 2011;50:601–3.
5. Andersen LK, Hercogova J, Wollina U, Davis MD. Climate change and skin disease: a review of the English-language literature. Int J Dermatol. 2012;51:656–61. quiz 659, 661
6. Grover S. Rajeshwari: global warming and its impact on skin disorders. Indian J Dermatol Venereol Leprol. 2009;75:337–9.
7. Balato N, Ayala F, Megna M, Balato A, Patruno C. Climate change and skin. G Ital Dermatol Venereol. 2013;148:135–46.
8. Vachiramon V, Busaracome P, Chongtrakool P, Puavilai S. Skin diseases during floods in Thailand. J Med Assoc Thail. 2008;91:479–84.
9. Lee SH, Choi CP, Eun HC, Kwon OS. Skin problems after a tsunami. J Eur Acad Dermatol Venereol. 2006;20:860–3.
10. Tak S, Bernard BP, Driscoll RJ, Dowell CH. Floodwater exposure and the related health symptoms among firefighters in New Orleans, Louisiana 2005. Am J Ind Med. 2007;50:377–82.
11. Swygard H, Stafford RE. Effects on health of volunteers deployed during a disaster. Am Surg. 2009;75:747–52. discussion 752–743
12. Appelgren P, Farnebo F, Dotevall L, Studahl M, Jonsson B, Petrini B. Late-onset posttraumatic skin and soft-tissue infections caused by rapid-growing mycobacteria in tsunami survivors. Clin Infect Dis. 2008;47:e11–6.
13. Hiransuthikul N, Tantisiriwat W, Lertutsahakul K, Vibhagool A, Boonma P. Skin and soft-tissue infections among tsunami survivors in southern Thailand. Clin Infect Dis. 2005;41:e93–6.

14. Svensson E, Welinder-Olsson C, Claesson BA, Studahl M. Cutaneous melioidosis in a Swedish tourist after the tsunami in 2004. Scand J Infect Dis. 2006;38:71–4.
15. Centers for Disease Control and Prevention (CDC). Infectious disease and dermatologic conditions in evacuees and rescue workers after hurricane katrina-multiple states, August–September, 2005. MMWR Morb Mortal Wkly Rep. 2005;54:961–4.
16. Stewart JH, Goodman MM. Earthquake urticaria. Cutis. 1989;43:340.
17. Gupta MA, Gupta AK. Psychodermatology: an update. J Am Acad Dermatol. 1996;34:1030–46.
18. Leclerc H, Schwartzbrod L, Dei-Cas E. Microbial agents associated with waterborne diseases. Crit Rev Microbiol. 2002;28:371–409.
19. Jansen A, Stark K, Schneider T, Schoneberg I. Sex differences in clinical leptospirosis in Germany: 1997–2005. Clin Infect Dis. 2007;44:e69–72.
20. Thaipadungpanit J, Wuthiekanun V, Chantratita N, Yimsamran S, Amornchai P, Boonsilp S, Maneeboonyang W, Tharnpoophasiam P, Saiprom N, Mahakunkijcharoen Y, Day NP, Singhasivanon P, Peacock SJ, Limmathurotsakul D. Leptospira species in floodwater during the 2011 floods in the Bangkok Metropolitan Region, Thailand. Am J Trop Med Hyg. 2013;89:794–6.
21. Agampodi SB, Dahanayaka NJ, Bandaranayaka AK, Perera M, Priyankara S, Weerawansa P, Matthias MA, Vinetz JM. Regional differences of leptospirosis in Sri Lanka: observations from a flood-associated outbreak in 2011. PLoS Negl Trop Dis. 2014;8:e2626.
22. Smith JK, Young MM, Wilson KL, Craig SB. Leptospirosis following a major flood in central Queensland, Australia. Epidemiol Infect. 2013;141: 585–90.
23. Amilasan AS, Ujiie M, Suzuki M, Salva E, Belo MC, Koizumi N, Yoshimatsu K, Schmidt WP, Marte S, Dimaano EM, Villarama JB, Ariyoshi K. Outbreak of leptospirosis after flood, the Philippines, 2009. Emerg Infect Dis. 2012;18:91–4.
24. Dechet AM, Parsons M, Rambaran M, Mohamed-Rambaran P, Florendo-Cumbermack A, Persaud S, Baboolal S, Ari MD, Shadomy SV, Zaki SR, Paddock CD, Clark TA, Harris L, Lyon D, Mintz ED. Leptospirosis outbreak following severe flooding: a rapid assessment and mass prophylaxis campaign; Guyana, January–February 2005. PLoS One. 2012;7:e39672.
25. Horak P, Mikes L, Lichtenbergova L, Skala V, Soldanova M, Brant SV. Avian schistosomes and outbreaks of cercarial dermatitis. Clin Microbiol Rev. 2015;28:165–90.
26. Wu XH, Zhang SQ, Xu XJ, Huang YX, Steinmann P, Utzinger J, Wang TP, Xu J, Zheng J, Zhou XN. Effect of floods on the transmission of schistosomiasis in the Yangtze river valley, people's Republic of China. Parasitol Int. 2008;57:271–6.
27. Barbosa CS, Leal-Neto OB, Gomes EC, Araujo KC, Domingues AL. The endemisation of schistosomiasis

in porto de galinhas, pernambuco, Brazil, 10 years after the first epidemic outbreak. Mem Inst Oswaldo Cruz. 2011;106:878–83.

28. Zhou XN, Yang GJ, Yang K, Wang XH, Hong QB, Sun LP, Malone JB, Kristensen TK, Bergquist NR, Utzinger J. Potential impact of climate change on schistosomiasis transmission in China. Am J Trop Med Hyg. 2008;78:188–94.

29. Khormi HM, Kumar L. Climate change and the potential global distribution of Aedes aegypti: spatial modelling using GIS and CLIMEX. Geospat Health. 2014;8:405–15.

30. Mourya DT, Yadav P, Mishra AC. Effect of temperature stress on immature stages and susceptibility of Aedes aegypti mosquitoes to chikungunya virus. Am J Trop Med Hyg. 2004;70:346–50.

31. Gilbert L. Altitudinal patterns of tick and host abundance: a potential role for climate change in regulating tick-borne diseases? Oecologia. 2010;162:217–25.

32. Roy-Dufresne E, Logan T, Simon JA, Chmura GL, Millien V. Poleward expansion of the white-footed mouse (Peromyscus leucopus) under climate change: implications for the spread of lyme disease. PLoS One. 2013;8:e80724.

33. Hlavacova J, Votypka J, Volf P. The effect of temperature on leishmania (kinetoplastida: trypanosomatidae) development in sand flies. J Med Entomol. 2013;50:955–8.

34. Centers for disease control and prevention. http://healthmap.Org/dengue/en/. Accessed 8 Jan 2015.

35. Bolivar-Mejia A, Alarcon-Olave C, Rodriguez-Morales AJ. Skin manifestations of arthropod-borne infection in Latin America. Curr Opin Infect Dis. 2014;27:288–94.

36. Bouzid M, Colon-Gonzalez FJ, Lung T, Lake IR, Hunter PR. Climate change and the emergence of vector-borne diseases in Europe: case study of dengue fever. BMC Public Health. 2014;14:781.

37. Sang S, Yin W, Bi P, Zhang H, Wang C, Liu X, Chen B, Yang W, Liu Q. Predicting local dengue transmission in Guangzhou, China, through the influence of imported cases, mosquito density and climate variability. PLoS One. 2014;9:e102755.

38. Banu S, Hu W, Guo Y, Hurst C, Tong S. Projecting the impact of climate change on dengue transmission in Dhaka, Bangladesh. Environ Int. 2014;63:137–42.

39. Hii YL, Rocklov J, Ng N, Tang CS, Pang FY, Sauerborn R. Climate variability and increase in intensity and magnitude of dengue incidence in Singapore. Glob Health Action. 2009;2 doi:10.3402/gha.v2i0.2036.

40. Williams CR, Mincham G, Ritchie SA, Viennet E, Harley D. Bionomic response of Aedes aegypti to two future climate change scenarios in far north Queensland, Australia: implications for dengue outbreaks. Parasit Vectors. 2014;7:447.

41. Tilley PA, Fox JD, Jayaraman GC, Preiksaitis JK. Maculopapular rash and tremor are associated with West Nile fever and neurological syndromes. J Neurol Neurosurg Psychiatry. 2007;78:529–31.

42. Paz S, Semenza JC. Environmental drivers of West Nile fever epidemiology in Europe and Western Asia--a review. Int J Environ Res Public Health. 2013;10:3543–62.

43. Wang G, Minnis RB, Belant JL, Wax CL. Dry weather induces outbreaks of human West Nile virus infections. BMC Infect Dis. 2010;10:38.

44. Harrigan RJ, Thomassen HA, Buermann W, Smith TB. A continental risk assessment of West Nile virus under climate change. Glob Chang Biol. 2014;20:2417–25.

45. Chen CC, Jenkins E, Epp T, Waldner C, Curry PS, Soos C. Climate change and West Nile virus in a highly endemic region of North America. Int J Environ Res Public Health. 2013;10:3052–71.

46. Soverow JE, Wellenius GA, Fisman DN, Mittleman MA. Infectious disease in a warming world: how weather influenced West Nile virus in the United States (2001–2005). Environ Health Perspect. 2009;117:1049–52.

47. Johnson BJ, Sukhdeo MV. Drought-induced amplification of local and regional West Nile virus infection rates in New Jersey. J Med Entomol. 2013;50:195–204.

48. Centers for disease control and prevention. www.Cdc.Gov/chikungunya/geo/index.Html. Accessed 5 Jan 2015.

49. Riyaz N, Riyaz A, Abdul Latheef EN, Anitha PM, Aravindan KP, Nair AS, Shameera P. Cutaneous manifestations of chikungunya during a recent epidemic in Calicut, North Kerala, South India. Indian J Dermatol Venereol Leprol. 2010;76:671–6.

50. Dommar CJ, Lowe R, Robinson M, Rodo X. An agent-based model driven by tropical rainfall to understand the spatio-temporal heterogeneity of a chikungunya outbreak. Acta Trop. 2014;129:61–73.

51. Ditsuwan T, Liabsuetrakul T, Chongsuvivatwong V, Thammapalo S, McNeil E. Assessing the spreading patterns of dengue infection and chikungunya fever outbreaks in lower southern Thailand using a geographic information system. Ann Epidemiol. 2011;21:253–61.

52. Chretien JP, Anyamba A, Bedno SA, Breiman RF, Sang R, Sergon K, Powers AM, Onyango CO, Small J, Tucker CJ, Linthicum KJ. Drought-associated chikungunya emergence along coastal East Africa. Am J Trop Med Hyg. 2007;76:405–7.

53. Anderson SG, French EL. An epidemic exanthem associated with polyarthritis in the Murray valley, 1956. Med J Aust. 1957;44:113–7.

54. Tong S, Hu W, McMichael AJ. Climate variability and Ross River virus transmission in Townsville Region, Australia, 1985–1996. Tropical Med Int Health. 2004;9:298–304.

55. Bi P, Hiller JE, Cameron AS, Zhang Y, Givney R. Climate variability and Ross River virus infections in Riverland, South Australia, 1992–2004. Epidemiol Infect. 2009;137:1486–93.

56. Tomerini DM, Dale PE, Sipe N. Does mosquito control have an effect on mosquito-borne disease? The case of Ross River virus disease and mosquito

management in Queensland, Australia. J Am Mosq Control Assoc. 2011;27:39–44.

57. Garza M, Feria Arroyo TP, Casillas EA, Sanchez-Cordero V, Rivaldi CL, Sarkar S. Projected future distributions of vectors of trypanosoma cruzi in North America under climate change scenarios. PLoS Negl Trop Dis. 2014;8:e2818.

58. Costa J, Dornak LL, Almeida CE, Peterson AT. Distributional potential of the triatoma brasiliensis species complex at present and under scenarios of future climate conditions. Parasit Vectors. 2014;7:238.

59. Cordovez JM, Rendon LM, Gonzalez C, Guhl F. Using the basic reproduction number to assess the effects of climate change in the risk of chagas disease transmission in Colombia. Acta Trop. 2014;129:74–82.

60. Guzman-Tapia Y, Ramirez-Sierra MJ, Escobedo-Ortegon J, Dumonteil E. Effect of hurricane isidore on triatoma dimidiata distribution and chagas disease transmission risk in the Yucatan Peninsula of Mexico. Am J Trop Med Hyg. 2005;73:1019–25.

61. Asano S, Mori K, Yamazaki K, Sata T, Kanno T, Sato Y, Kojima M, Fujita H, Akaike Y, Wakasa H. Temporal differences of onset between primary skin lesions and regional lymph node lesions for tularemia in japan: a clinicopathologic and immunohistochemical study of 19 skin cases and 54 lymph node cases. Virchows Arch. 2012;460:651–8.

62. Nakazawa Y, Williams R, Peterson AT, Mead P, Staples E, Gage KL. Climate change effects on plague and tularemia in the United States. Vector Borne Zoonotic Dis. 2007;7:529–40.

63. Balci E, Borlu A, Kilic AU, Demiraslan H, Oksuzkaya A, Doganay M. Tularemia outbreaks in kayseri, turkey: an evaluation of the effect of climate change and climate variability on tularemia outbreaks. J Infect Public Health. 2014;7:125–32.

64. Ryden P, Sjostedt A, Johansson A. Effects of climate change on tularaemia disease activity in Sweden. Glob Health Action. 2009;2 doi:10.3402/gha.v2i0.2063.

65. Vig DK, Wolgemuth CW. Spatiotemporal evolution of erythema migrans, the hallmark rash of lyme disease. Biophys J. 2014;106:763–8.

66. Jaenson TG, Lindgren E. The range of ixodes ricinus and the risk of contracting lyme borreliosis will increase northwards when the vegetation period becomes longer. Ticks Tick Borne Dis. 2011;2:44–9.

67. Brownstein JS, Holford TR, Fish D. Effect of climate change on lyme disease risk in North America. EcoHealth. 2005;2:38–46.

68. Tuite AR, Greer AL, Fisman DN. Effect of latitude on the rate of change in incidence of lyme disease in the United States. CMAJ Open. 2013;1:E43–7.

69. Ogden NH, Maarouf A, Barker IK, Bigras-Poulin M, Lindsay LR, Morshed MG, O'Callaghan CJ, Ramay F, Waltner-Toews D, Charron DF. Climate change and the potential for range expansion of the lyme disease vector ixodes scapularis in Canada. Int J Parasitol. 2006;36:63–70.

70. Trajer A, Bobvos J, Paldy A, Krisztalovics K. Association between incidence of lyme disease and spring-early summer season temperature changes in Hungary—1998–2010. Ann Agric Environ Med. 2013;20:245–51.

71. World Health Organization. http://apps.Who.Int/neglected_diseases/ntddata/leishmaniasis/leishmaniasis.Html. Accessed 8 Jan 2015.

72. Gonzalez C, Wang O, Strutz SE, Gonzalez-Salazar C, Sanchez-Cordero V, Sarkar S. Climate change and risk of leishmaniasis in North America: predictions from ecological niche models of vector and reservoir species. PLoS Negl Trop Dis. 2010;4:e585.

73. Peterson AT, Shaw J. Lutzomyia vectors for cutaneous leishmaniasis in southern Brazil: ecological niche models, predicted geographic distributions, and climate change effects. Int J Parasitol. 2003;33:919–31.

74. Bounoua L, Kahime K, Houti L, Blakey T, Ebi KL, Zhang P, Imhoff ML, Thome KJ, Dudek C, Sahabi SA, Messouli M, Makhlouf B, El Laamrani A, Boumezzough A. Linking climate to incidence of zoonotic cutaneous leishmaniasis (L. Major) in Pre-Saharan North Africa. Int J Environ Res Public Health. 2013;10:3172–91.

75. Fischer D, Moeller P, Thomas SM, Naucke TJ, Beierkuhnlein C. Combining climatic projections and dispersal ability: a method for estimating the responses of sandfly vector species to climate change. PLoS Negl Trop Dis. 2011;5:e1407.

76. Galvez R, Descalzo MA, Guerrero I, Miro G, Molina R. Mapping the current distribution and predicted spread of the leishmaniosis sand fly vector in the Madrid region (Spain) based on environmental variables and expected climate change. Vector Borne Zoonotic Dis. 2011;11:799–806.

77. Jafari N, Shahsanai A, Memarzadeh M, Loghmani A. Prevention of communicable diseases after disaster: a review. J Res Med Sci. 2011;16:956–62.

Modern Skin Toxicity Testing Strategies

4

Susanne N. Kolle, Wera Teubner, and Robert Landsiedel

4.1 Introduction

Substances may affect the skin in many ways, which may result in a variety of adverse effects. Clinically observable pathological skin conditions are as manifold as their underlying mechanisms. Local effects on the skin include corrosion, irritation and sensitization. *Corrosion* is defined as an irreversible destruction of the skin, whereas *irritation* implies reversible damage that is mostly caused by inflammation. However, substances may not only induce inflammation directly (i.e. by irritation) but also by immune-mediated processes (i.e. by sensitization). Allergic contact dermatitis is the clinical manifestation of *skin sensitization*, i.e. the immune-mediated inflammation of the skin. It requires repeated dermal contact with the substance, whereas direct irritation and corrosion are typically acute effects that are already observed after one single contact. Additionally, some substances may elicit local dermal effects only upon simultaneous irradiation with UV or visible light, thereby causing either *phototoxicity* (photoirritation or, in vitro, photo-cytotoxicity) or *photoallergic reactions* (with the frequently used but misleading overarching term for all UV- and visible light-facilitated effects, 'photosensitivity'; [1]).

The present chapter focuses on modern testing strategies for local effects resulting from dermal contact to substances. 'Modern testing strategies' aim at collecting mechanistically relevant toxicological information. Studies are designed making use of the abundance of toxicological in vivo data already available in large databases, also by computational processing. The mechanistically relevant information is collected and evaluated in integrated approaches for the testing and assessment of substances (IATAs) that take into account extensive physico-chemical characterization of the test substances and make use of in vitro test methods and in silico models (see below) for testing and only refer to in vivo studies as a last resort. The mechanistically relevant toxicological information allows the linking of molecular initiating events to resulting apical toxic effects as described in the Adverse Outcome Pathway (AOP) concept [2]. Furthermore, modern testing strategies contribute to refining, reducing or even replacing animal testing in line with the 3Rs principle [3] that has been implemented in European Union (EU) legislation [4].

The present chapter does not cover systemic effects of dermally applied substances or skin toxicity arising upon substance ingestion (e.g. chloracne). In brief, such effects may be identified using internationally harmonized test

S.N. Kolle • R. Landsiedel (✉)
BASF SE, Experimental Toxicology and Ecology,
Ludwigshafen am Rhein, Germany
e-mail: robert.landsiedel@basf.com

W. Teubner
BASF Schweiz AG, Product Safety,
Basel, Switzerland

© Springer International Publishing Switzerland 2018
J. Krutmann, H.F. Merk (eds.), *Environment and Skin*,
https://doi.org/10.1007/978-3-319-43102-4_4

methods (e.g. the Organisation for Economic Co-Operation and Development (OECD) test guidelines (TGs) 402, 410 and 411 for dermal toxicity tests and OECD TGs 407 and 408 for oral toxicity tests). Of note, all OECD TGs are available at http://www.oecd.org/env/ehs/testing/oecdguidelinesforthetestingofchemicals.htm. Genotoxic effects in the skin are also excluded from the scope of the present chapter. Generally, genotoxicity is assessed using in vitro and in vivo methods (e.g. OECD TGs 471, 473, 474, 487, 489), which are not skin specific. However, the development of genotoxicity tests using reconstructed skin models is under way [5].

In most substance or product safety legislations, the identification of potential hazards of substances arising from incidental skin contact or intended topical application is a core requirement. The *Globally Harmonized System of Classification and Labelling of Chemicals* (GHS; [6]) that has been implemented—albeit with modifications—in the *EU Regulation on the classification, labelling and packaging of substances and mixtures* [7] distinguishes three types of local dermal effects with the following categories and hazard phrases: skin corrosion (Category 1, corrosion H314), skin irritation (Category 2, irritation H315, and Category 3, mild irritation H316) and skin sensitization (Category 1, sensitization H317). Additionally, *repeated exposure may cause skin dryness or cracking* (EUH066) has been laid down in the EU as further skin-related hazard phrase [8]. All hazard categories may be sub-categorized by potency levels (e.g. Sub-categories 1A, 1B and 1C for 'Category 1, corrosion', and Sub-categories 1A and 1B for 'Category 1, sensitization', with the latter sub-categorization based upon human evidence or the potency of effects observed in animals). Notwithstanding, the GHS does not require sub-categorization, and it may indeed not improve the ensuing risk management measures. Specific GHS hazard phrases to indicate phototoxic potential or potency are not available. For substances, the GHS entered into force in the EU in 2010 and was implemented for mixtures in mid-2015.

In 'traditional' local dermal toxicity tests, test substances are applied onto the skin of animals (once for irritation and corrosion testing and repeatedly for sensitization testing), and the resulting reactions are observed. Depending on the type of substance under investigation and the corresponding legislative requirements, the traditional animal tests are still mandatory. By contrast, modern toxicological methods are not restricted to observing the clinical outcome of the exposure to a substance. Instead, they also aim at elucidating the specific underlying mechanisms of toxicity relevant for humans. One of the first methods rendering this possible for skin sensitization was an in vivo test, the murine local lymph node assay (LLNA, OECD TG 429). Instead of inspecting the outbreak (elicitation phase) of an allergic contact dermatitis in guinea pigs, the LLNA (using mice) analyses the process of sensitization (induction phase) without actually provoking an outbreak of the disease.

Increasingly, in vitro test methods are gaining importance to investigate mechanisms of toxicity. In addition to (and also as a consequence of) the focus on revealing toxicological mechanisms in humans, many modern in vitro test methods utilize cultured cells and tissues of human origin. Thereby, the scientific limitations of extrapolating effects in animals to effects in humans are avoided, which is expected to further reduce and ultimately replace the need for animal testing for local toxicity assessments. One of the first toxicological in vitro methods adopted by the OECD in 2004 (and revised in 2014, 2015 and 2016) was the skin corrosion test (OECD TG 431) using reconstructed human epidermis (RhE) models (cf. Sect. 4.2). Since the adoption of OECD TG 431 more than 10 years ago, also in the areas of phototoxicity testing (presented in Sect. 4.3) and skin sensitization testing (presented in Sect. 4.4), a number of in vitro assays have been developed and validated, and their regulatory acceptance has either been finalized or is currently ongoing.

In silico models are further indispensible components of modern toxicity testing strategies. These models include (quantitative) structure activity relationships ((Q)SARs) that use computational methods to evaluate the physico-chemical or structural properties (descriptors) of a molecule to assign it to a certain category or biological activity. Such modelling may be based upon different elements, depending on the quantity, quality and detail of experimental data available and the extent of understanding of the underlying mechanisms. (Q)SAR models may be used as stand-alone methods aiming at screening for adverse properties, but they may have even greater benefit when applied in combination with other methods of modern toxicology, thereby aiming at replacing the need for experimental testing. In accordance with Annex XI of the *EU Regulation for the registration, evaluation, authorisation and restriction of chemicals* (REACH, [9]), (Q)SAR screening for adverse properties of chemicals may serve to justify waiving of testing needs during hazard assessment by read-across, category or weight-of-evidence (WoE) approaches.

To ensure transparency on (Q)SAR models and their appropriate regulatory use, the OECD has established five so-called principles for the characterization and documentation of (Q)SAR models [10]. The *first principle* requests associating every (Q)SAR with a defined endpoint. This enforces that the experimental data underlying the model is critically evaluated specifying what the model is intended to predict. For instance, a number of skin irritation models are based upon the so-called primary irritation index (PII), which is calculated from a grading of the severity of dermal effects during the first 3 days after substance exposure. Therefore, PII scores do not include information on the reversibility of effects, which impairs the applicability of PII-based models for GHS classification and labelling. The *second and third principles* require expressing (Q)SARs in the form of defined algorithms and associating them with defined applicability domains. Models developed for a certain type of chemical class may perform well for this specific class but may not be suitable for other classes of substances. Hence, the applicability domain provides an indication on the reliability of a given predication. In practice, applicability domains are often defined with respect to certain mechanistic and/or statistic parameters, so that an overall assessment of a model's reliability may be not feasible. Several models provide information on domain adherence for

known parameters and highlight all unknown fragments (of the respective test substance). Therefore, even though a substance may be assignable to the respective domain of a given model, the prediction of its effects will not necessarily be reliable, unless the overall domain definition is so strict that it does not allow for unknown features. Such aspects of (Q)SAR characterization reflect limitations in regard to the size of the training set or to the respective mechanistic understanding. The *fourth principle* encompasses requirements on how to describe the model's performance using the training set substances (internal validation) and unknown substances (external validation). Details of these requirements depend on whether the given model is intended for screening purposes or for actual hazard assessments. Finally, providing a mechanistic explanation of the model is the content of the *fifth principle* [10].

4.2 Skin Irritation and Corrosion Testing

Recently, Worth et al. [11] summarized the mechanistic knowledge on skin irritation and corrosion: corrosion may occur if inorganic acids and bases and strong organic acids with extreme pH erode the *stratum corneum*. For this reason, substances with a pH value below 2 or above 11.5 may be classified as corrosives without further testing. In addition, certain substance classes known to react with skin components may damage the skin, especially if their surface activity is high and they penetrate the skin easily. For skin irritation, Worth et al. [11] reported that a putative AOP is under development by the European Commission's Joint Research Centre. This putative AOP is based on the following key events: dermal bioavailability and damage to the dermal barrier, metabolism, chemically induced tissue trauma, release of inflammatory mediators and activation of the innate immune system. Nevertheless, Worth and co-authors caution that dermal irritation is not necessarily mechanistically fully understood; as of April 2017, no AOP for skin irritation has been published.

In respect to in vivo testing, the Draize skin irritation and corrosion test has been used for many decades to predict acute skin irritation and corrosion hazard ([12]; OECD TG 404). This 'traditional' animal test implies single topical application of liquid or solid test substances (via gauze patches) onto the intact skin of rabbits. After 4 h exposure, the patches (and test substance residuals on the skin) are removed. Cutaneous reactions are assessed immediately after removal of the patch and at specific intervals during 14 days post-exposure. Skin reactions are evaluated by grading erythema as well as eschar and oedema formation, assessing each of the latter individually. For many years, the Draize skin irritation test has been subject to profound criticism. Besides animal welfare concerns, the test is highly variable [13] and over-predictive of in vivo irritation responses in humans [14–16]. To reduce and refine in vivo skin irritation and corrosion testing, a sequential procedure is described in a supplement to OECD TG 404. Accordingly, the test substance is initially applied to one single animal. Two further animals are only used if skin corrosivity is not observed in the first animal.

In addition to the RhE model-based skin corrosion test (OECD TG 431), in 2010, the corresponding skin irritation test (OECD TG 439), which also utilizes RhE models, was adopted (with revisions in 2013 and 2015). RhE models, which are relatively easy to handle and use, are commercially available in different regions of the world (e.g. [17–22]). RhE models closely mimic the biochemical and physiological properties of the human epidermis. They further mimic the cell and tissue damage resulting in localized trauma that is the underlying mechanism of in vivo irritation [13]. In the RhE-based tests, relative viability is determined after test substance exposure using the 3-(4,5-dimethylthiazol-2-yl)-2,5-diphenyltetrazolium bromide (MTT) assay.

Currently, both OECD TG 439 and OECD TG 431 encompass four similar but distinct test methods, which use different commercially available RhE models as well as different model-specific test protocols and prediction models. All four RhE-based skin corrosion tests have been adopted for substance assignment to the sub-categories of the GHS Category 1 corrosion, i.e. Sub-categories 1A and 1B/1C combined, and 'non-corrosivity' (UN, 2013). However, as described in further detail in OECD TG 431 (revision 2013), depending on the specific test used, up to 46% of the Sub-category 1B/1C substances may be over-predicted as Sub-category 1A substances. By a revision of the prediction model in 2016, this over-prediction rate was reduced to about 30%. Regarding skin irritation, the test methods described in OECD TG 439 have been adopted to distinguish non-irritants from GHS Category 2 irritants. By contrast, they are not considered adequate for the identification of the optional GHS Category 3 mild irritants, which however is not required in all regulatory contexts. Test substances that are identified as irritants in accordance with OECD TG 439 should be subjected to additional testing to assess or rule out skin corrosivity.

Two further skin corrosion test guidelines are available in addition to the RhE-based tests. In the transcutaneous electrical resistance (TER) method (OECD TG 430), explanted rat skin discs are treated with the test substance, and skin integrity and barrier function are measured by means of TER. The test can be used to distinguish corrosives from non-corrosives but not to further assign corrosives to the GHS Sub-categories 1A, 1B and 1C. Of note, in some jurisdictions, this test is considered to be an animal test, and it is not further discussed in the present chapter. Currently, the only test that has been adopted for the full sub-categorization into the GHS Sub-categories 1A, 1B and 1C is the in vitro membrane barrier test (OECD TG 435). Basically, in this test the corrosivity of a substance is detected by a pH-indicator solution, and the time required for a test substance to break through a biomembrane (commercially available as Corrositex®) is determined. However, this test is limited to substances with proven compatibility with the pH-indicator solution.

In summary, as illustrated in Fig. 4.1, using a combination of the mentioned in vitro tests, the in vivo Draize skin irritation test may be fully replaced, unless assessment of GHS Category 3 (mild irritancy) is required. Test combinations are

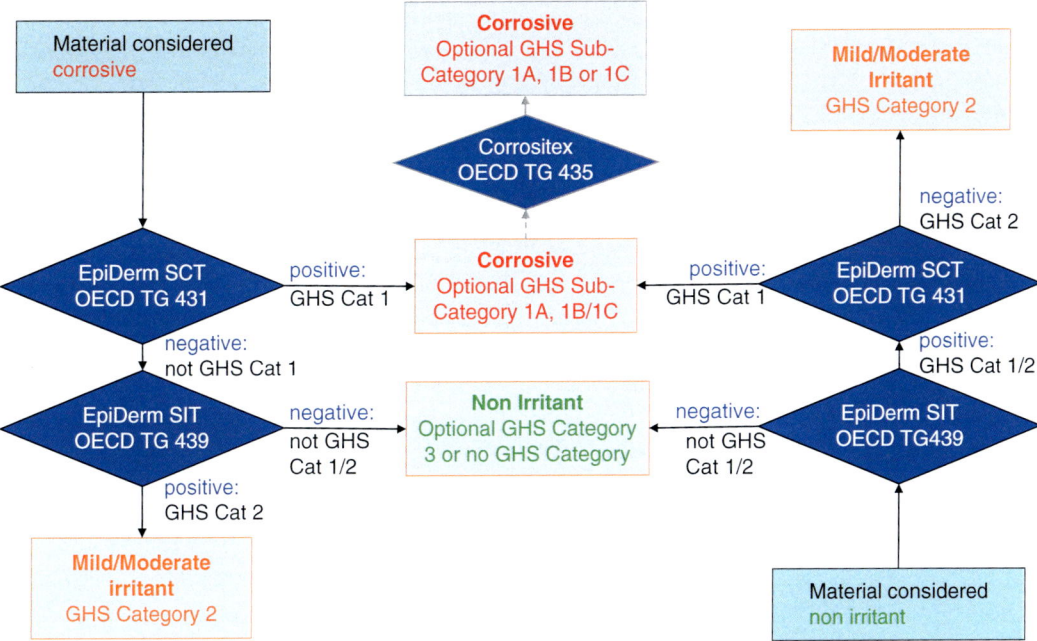

Fig. 4.1 Tiered in vitro skin irritation and corrosion testing using the OECD TGs 431, 439 and 435 to replace OECD TG 404. *Cat* category, *RhE* reconstructed human epidermis, *SCT* skin corrosion test, *SIT* skin irritation test

chosen depending on the expected irritating potential of the given test substance and on the possible need for sub-categorization of corrosivity.

In respect to in silico models, already more than 15 years ago, the German Federal Institute for Risk Assessment (BfR—*Bundesinstitut für Risikobewertung*) processed available mechanistic knowledge on local dermal toxicity in a decision support system (DSS) for the classification of substances as corrosives, irritants or non-irritants [23, 24]. The DSS contains a number of physico-chemical *exclusion rules* (addressing molecular weight, water solubility, fat solubility, n-octanol-water partition coefficient, vapour pressure and surface tension) and a set of structural alerts (*inclusion rules*). However, the DSS is based on the classification scheme used in the EU before the GHS first entered into force in 2010, and both schemes differ in respect to the definitions for different levels of irritancy. As a result, the physico-chemical exclusion rules of the DSS are not suitable to assess mild irritants as defined by the GHS Category 3.

Most inclusion and exclusion rules of the DSS only allow distinguishing corrosives from non-corrosives, and they do not allow assessing all types of substances. For example, the parameter 'surface tension' is only applicable for molecules with formulas of $C_xH_yO_z$ or $C_xH_yO_zN_aS_b$. For these molecules, the prediction 'not corrosive' (i.e. not Category 1 or not R34/R35 in accordance with the previous classification scheme) is made if their surface tension exceeds 62 mN/m, which is the threshold value to distinguish surface-active from non-surface-active substances. Information on (non-)irritancy cannot be obtained from this parameter alone. Among those substances to which all *exclusion rules* apply, non-corrosives and non-irritants are usually recognized with high predictivity [25]. The structural alerts (*inclusion rules*) perform well to identify corrosives, but they only have a moderate predictivity (68%) for irritants [26].

Both rule sets have been included in the OECD QSAR Toolbox (www.qsartoolbox.org) and in Toxtree (http://toxtree.sourceforge.net). In the OECD QSAR Toolbox, they are termed profilers and serve to support a mechanistic justification of read-across, trend analysis or substance-tailored QSAR. Many commercial

QSAR model providers offer modules for the prediction of skin irritation. In a comparison of the DEREK, HAZARDEXPERT and TOPKAT QSAR software packages, only TOPKAT was able to predict the irritation potential for the majority of chemicals tested [27]. The OECD *guidance document on an integrated approach on testing and assessment (IATA) for skin corrosion and irritation* [28] contains an overview of published models that may exploit the large database of existing experimental data on skin irritation and corrosion. Overall, skin irritation and corrosion, especially when based on pH or other chemical interactions, may well be predicted by simple rules or more comprehensive in silico models.

4.3   Phototoxicity Testing

Substances that are applied onto sun-exposed skin (or that come into contact with it incidentally) and that absorb sunlight (typically at wavelengths of 290–700 nm) may form active molecules, which cause skin irritation or sensitization, while the nonirradiated substances alone do not—or only to a lesser extent [29]. Such active molecules may be formed by irradiation either 'non-photodynamically' from the substances themselves or 'photodynamically' by reactions of the substances with the oxygen present in the air [30]. In consequence, phototoxic or photoallergic effects may evolve [1]; however, most assessments focus on phototoxicity.

In vivo phototoxicity test methods using guinea pigs, rats or mice are available [31], but they have not yet been formally validated [32, 33]. By contrast, for in vitro phototoxicity testing, a standardized and internationally harmonized method is available, i.e. the in vitro 3T3 Neutral Red Uptake (NRU) phototoxicity test method (OECD TG 432; [34]). It measures the cytotoxicity (NRU) of a test substance on monolayers of immortalized fibroblasts (Balb/c 3T3 cells) in the presence and absence of light. To evaluate test results, the complete concentration-response curves obtained upon test substance incubation with and without light are analysed.

Accounting for the complexity of the data evaluation procedure, this is conducted using software provided by the OECD. The in vitro 3T3 NRU phototoxicity test was found to be over-predictive when compared to phototoxic effects elicited in human volunteers [35]. Additionally, test substances should be soluble in the aqueous culture media, as is also the case for most other cell-based in vitro assays. To overcome this limitation, an in vitro phototoxicity test method using the RhE model EpiDerm™ was developed and successfully pre-validated [36]. Nevertheless, more than 15 years later, an internationally accepted guideline for such a RhE-based phototoxicity test is unavailable. Just as the in vitro 3T3 NRU phototoxicity test, also the RhE-based assay was found to be over-predictive of phototoxicity in humans. Three out of four bergamot oils tested positive in vitro, whereas only two of these oils elicited effects in the photopatch test conducted with human volunteers [32].

Since many phototoxic substances act photodynamically and form reactive oxygen species (ROS) when irradiated, assays measuring ROS formation have been proposed for the assessment of phototoxicity. An intra- and interlaboratory validation study confirmed the reliability and relevance of such a ROS assay [37]. An OECD draft test guideline is available since in fall 2016 [38].

Likewise, in silico models to predict phototoxicity focus on ROS generation following light absorption. The HOMO-LUMO gap (an indicator of the photoreactivity of a substance) and/or classical molecular descriptors have been used as a basis for such in silico models [39, 40].

Phototoxicity testing is conducted extensively in the pharmaceutical sector using a modified 3T3 NRU phototoxicity assay since the test method adopted by the OECD was not specifically designed or validated for pharmaceutical substances [31, 35]. A tiered testing strategy is presented in the Guideline S10 of the *International Conference on Harmonization of Technical Requirements for Registration of Pharmaceuticals for Human Use* ([31]; Fig. 4.2). The first step of this testing strategy implies assessing whether the test substance absorbs light between 290 and 700 nm. Since this process

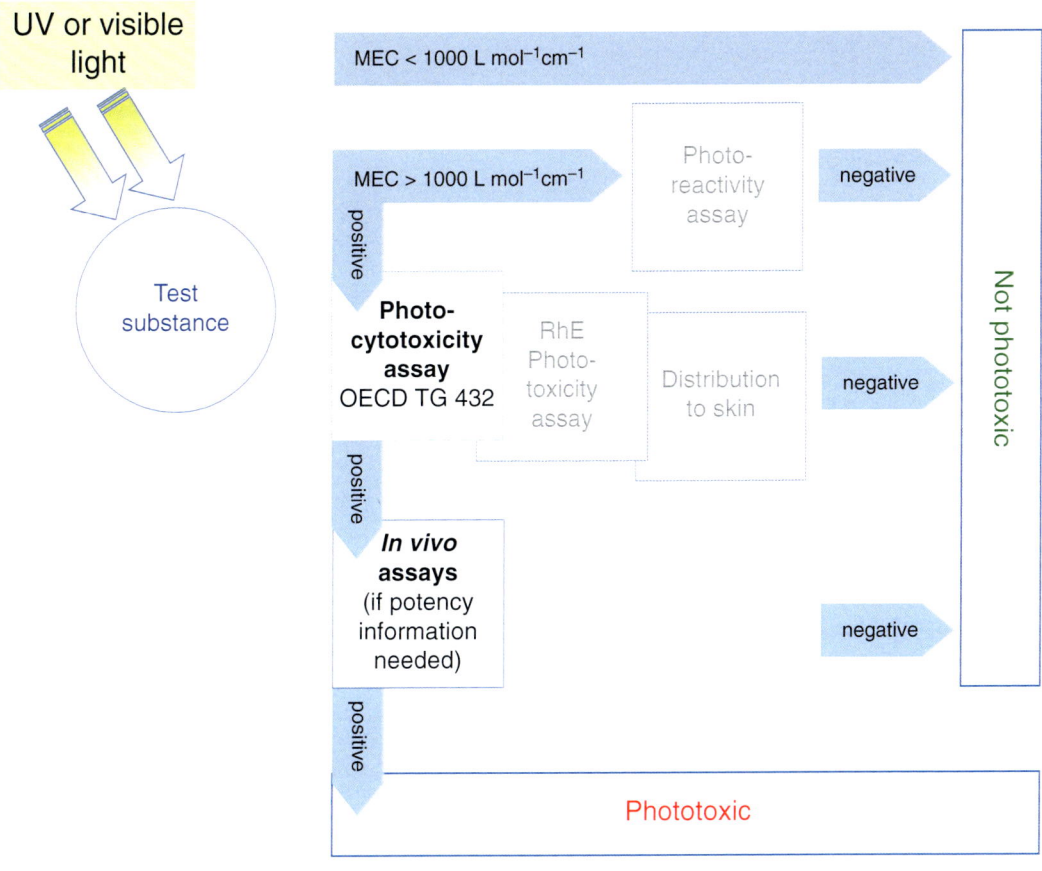

Fig. 4.2 Phototoxicity assessment strategies according to the ICH guideline S10 (simplified; [31]). *MEC* molar extinction coefficient, *RhE* reconstructed human epidermis

requires the transformation of energy from light to the molecules, substances with low molar extinction coefficient (i.e. <1000 L/[mol × cm]) do not have to be tested further. This property may be assessed spectrophotometrically, and for most substances, such a spectrophotometric test combined with in vitro photo-cytotoxicity testing, further taking into account exposure information, is sufficient for safety evaluation.

4.4 Skin Sensitization Testing

Skin sensitization and the outbreak of allergic contact dermatitis are complex biological processes involving interactions of the substance with skin proteins and interactions of different types of cells located in the skin and local lymph nodes. Key events of this process have been identified and put in order to describe an AOP for skin sensitization allowing anchoring of the molecular initiating event to the final apical toxic effect.

The four key events of the skin sensitization AOP are (1) covalent interaction with skin proteins, (2) events in keratinocytes, (3) events in dendritic cells and (4) events in lymphocytes, i.e. T cells ([41, 42]; Fig. 4.3): even though the *stratum corneum* of the skin is generally an effective barrier, specific substances that are typically of low molecular weight may be able to permeate into the skin. The molecular initiating event of skin sensitization occurs if such a substance (also called a 'hapten') modifies a skin protein. In many cases, an electrophilic hapten binds directly to a protein thereby forming an antigen (or 'allergen', first key event of the AOP). Alternatively,

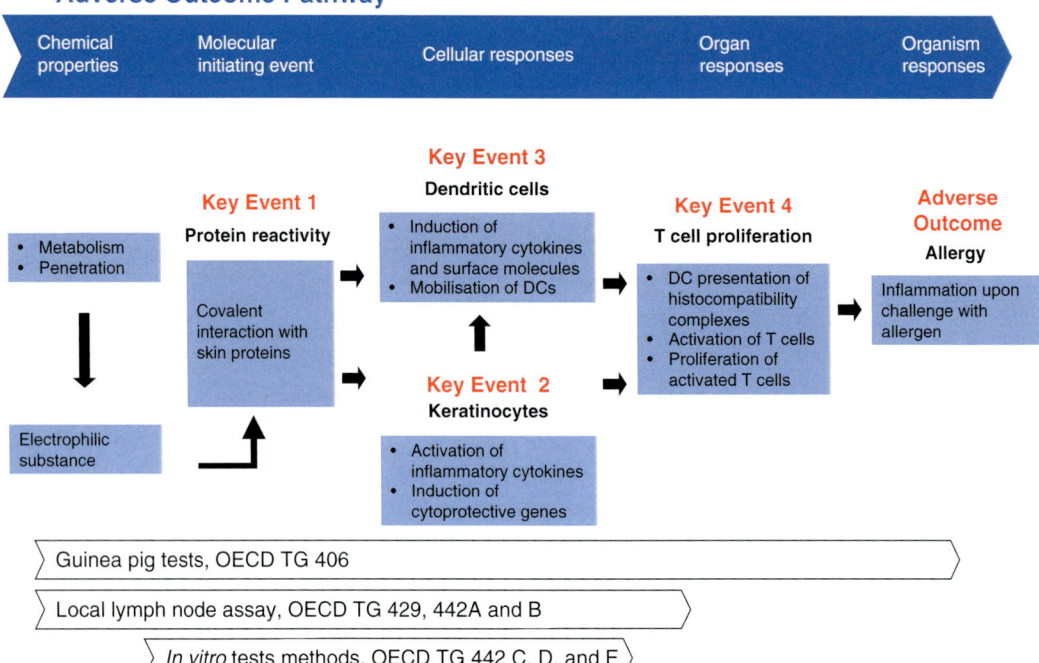

Fig. 4.3 The Adverse Outcome Pathway (AOP) of skin sensitization [41, 42]. Four key events of the AOP: (1) protein interaction, (2) keratinocyte activation, (3) dendritic cell (DC) activation and (4) T cell proliferation (adapted from [41]). The diagram also indicates which parts of the AOP are covered by the respective tests

the substance may be a pre- or pro-hapten and require transformation into an electrophilic hapten in order to induce the sensitization process. This transformation may either occur spontaneously in the skin or upon induction by dermal enzymes [43]. Subsequently, haptens must elicit the generation of 'danger signals' for *the immune system* to be *called into action* [44, 45]. This is initiated by the activation of keratinocytes (second key event). The third key event involves the activation of dendritic cells (DCs), such as Langerhans cells. DCs are motile cells, which take up the allergen in the skin and transfer it to the draining lymph nodes where they additionally activate T cells. In the course of this step, the epidermal DCs undergo a number of phenotypic and functional changes [46–48]. The fourth key event involves the proliferation of the allergen-specific T cells (*induction phase*) and the generation of allergen-specific memory T cells. The latter recognize subsequent contacts with the same hapten, upon which they trigger the inflam-

matory response of an allergic contact dermatitis (*elicitation phase*) [49].

Traditionally, evaluation of the sensitization potential of a substance has been carried out using animal studies, i.e. the guinea pig tests according to Magnusson and Kligman or Buehler (OECD TG 406). These tests cover the entire process of skin sensitization including the elicitation phase. Freund's complete adjuvant that may cause painful skin damage is used in the Magnusson and Kligman guinea pig maximization test but not in the Buehler test. By comparison, the murine LLNA (OECD TG 429; cf. Sect. 4.1) and its non-radioactive variants OECD TGs 442A and 442B are considered reduction and refinement methods. They use fewer animals than the guinea pig tests and inflict less distress on the animals, also by restricting the test method to the induction phase. Cell proliferation in the lymph nodes is quantified either by measuring incorporated radiolabelled thymidine or by non-radioactive methods (OECD TG 442As and B,

[50, 51]), making the LLNA and its variants less subjective than the guinea pig tests. The results of both guinea pig tests rely on a subjective read-out of skin reactions (erythema and oedema), and they provide limited information on the sensitizing potency of test substances.

The LLNA is accepted to evaluate the potency of sensitizing substances. Until recently the mouse based assay remained the preferred method for the assessment of chemicals [9]. However, just like any toxicological test method, the LLNA and its variants have scientific limitations. Certain substances and substance classes are known to produce false-positive [52–54] or false-negative results [51]. Additionally, and also in concordance with other 'biological' tests, the LLNA is subject to variability [55, 56]. This is particularly true for 'borderline' substances producing proliferative responses close to the threshold value for sensitizing substances [57, 58].

A multitude of in vitro assays has been developed to address the distinct key events described in the skin sensitization AOP, and there is widespread agreement that, presently, no single in vitro test is able to cover all steps leading to skin sensitization, let alone allergic contact dermatitis [59].

Several in vitro assays for skin sensitization testing have been validated and proven fit for regulatory purposes, and recently, three assays have even been adopted as OECD TGs. The OECD TG 442C, *In Chemico Skin Sensitisation: Direct Peptide Reactivity Assay (DPRA)*, addresses the first key event (protein reactivity). It foresees incubating the test substance with two model peptides (either a lysine- or a cysteine-containing heptapeptide). The amount of peptide remaining after test substance incubation is measured by high-performance liquid chromatography, and the fraction of the peptide that has been modified ('depleted') is calculated. The *Keratinocyte-based ARE-Nrf2 Luciferase Reporter Gene Test Method* (OECD TG 442D) addresses the second key event (keratinocyte activation). The KeratinoSens™ [60] and the LuSens assays [61, 62] are two test systems to investigate the keratinocyte ARE-Nrf2 (antioxidant response element-nuclear erythroid

2-related factor 2) pathway. Specifically, both assays reveal the activation of the Keap1-Nrf2 regulatory pathway [63]: electrophilic test substances that have been taken up into the keratinocytes may bind to the suppressor protein Keap1. This triggers the release of Nrf2 from the Keap1-Nrf2 complex, and Nrf2 may translocate into the cell nucleus. There it may bind to the ARE, which induces the expression of specific cytoprotective genes. The KeratinoSens™ and LuSens assays use immortalized human keratinocytes which have a reporter gene for luciferase that is controlled by ARE.

The *human Cell Line Activation Test* (h-CLAT) addresses the third key event (activation of DCs) and has been adopted as OECD TG 442E. The h-CLAT allows evaluating the activation of immortalized, human, dendritic-like THP-1 cells by expression of two surface markers, CD86 and CD54 [64, 65]. All of the mentioned in vitro methods use aqueous systems and are therefore generally limited to water-soluble test substances. Although all of the methods (may) provide continuous ranges of read-outs, currently, the OECD TGs only foresee their use to provide yes/no answers based upon predefined thresholds (i.e. to predict sensitization potential). The quantitative information they also provide is usually not utilized to predict sensitization potency (e.g. to assign substances to the GHS Sub-categories 1A and 1B).

The mentioned assays address individual key events of the AOP. The results of these assays (and any other relevant information) are combined to predict the skin sensitization potential. This implies that the results of the assays are evaluated jointly in a predefined way, using a prediction model, or in a not prescribed weight of evidence approach (WoE) [66]. Such approaches are termed integrated assessment and testing approaches (IATAs) and are currently being developed for different regulatory purposes [67]. Specifically, for the endpoint skin sensitization, 12 IATAs have been listed as case studies by the OECD [68]. These case studies present simple IATAs—like a '2 out of 3' prediction model—and complex Bayesian or artificial neural network models [69–71].

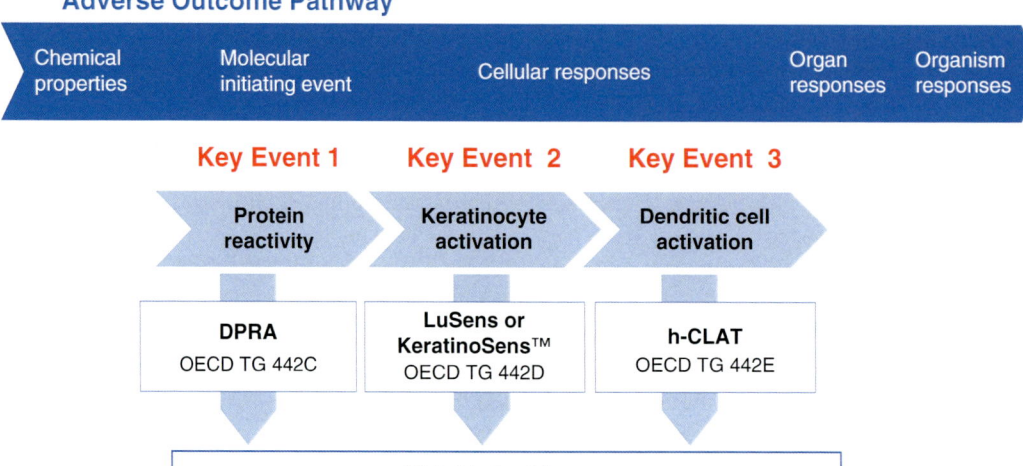

Fig. 4.4 Simple prediction model using modern testing methods addressing the early key events of the skin sensitization adverse outcome pathway: '2 out of 3' prediction model. *DPRA* direct peptide reactivity assay, *h-CLAT* human Cell Line Activation Test

The so-called 2 out of 3 prediction model (Fig. 4.4; [72]) uses the results of (at least) two of three assays (i.e. DPRA, KeratinoSens™/LuSens, h-CLAT) to define the overall prediction: if at least two assays are positive, the substance is considered to have sensitizing potential—or none, if at least two assays are negative. The '2 out of 3' prediction model has been applied to different data sets [72–74]. In all studies, it yielded higher accuracies than the individual assays, and it predicted effects in humans more reliably than LLNA data. Assessing more than 200 substances, Urbisch et al. [74] present an extensive evaluation of the '2 out of 3' prediction model which predicted the sensitization potential of 90% of the substances correctly. Based upon the protein-binding reaction chemistry of a substance, *mechanistic domains* may be defined for individual assays [75]; this offers a refined view on their applicability to predict skin sensitization potential in humans [74]. One question about the nonanimal methods is their ability to correctly predict pre-haptens and pro-haptens. Recently, the proficiency of the '2 out of 3' model to identify not only direct-acting haptens but also 22 of 27 pre-haptens and pro-haptens was demonstrated [76].

There is ample data to demonstrate the usability of nonanimal methods using IATAs. In Europe, the REACH legislation now prescribes the use of nonanimal methods over animal testing for skin sensitization potential.

While the skin sensitization potential can be reliably predicted by nonanimal approaches, methods to measure the skin sensitization potency are still in development, and the REACH legislation falls back to LLNA testing to discriminate the potency Sub-categories 1A and 1B.

In silico modelling of substance-induced skin sensitization benefits from the circumstance that the underlying mechanism of this effect is already known and an agreed AOP is available. Furthermore, the molecular initiating event of skin sensitization is not a biological process but a chemical one, i.e. the chemical modification of a dermal protein by the test substance. Chemical reactivity may be predicted by simple rules (reaction classes) or more comprehensive in silico models (usually with much higher accuracy than predictions of more complex and less understood biological processes).

(Q)SAR models for skin sensitization either include current mechanistic knowledge such as substance electrophilicity (e.g. DEREK, Toxtree, TIMES-SS), or they are exclusively based on statistical evaluations of structure fragments determined from the model training set (e.g. Case Ultra, Vega, TOPKAT; [77]). The OECD QSAR Toolbox provides mechanistic information along the skin sensitization AOP from which substance-specific (Q)SAR models may be created. Apart from a database of existing experimental data, the OECD (Q)SAR Toolbox contains two so-called profilers that scan chemical structures for structural alerts for protein binding. These profilers may be combined with simulators of skin metabolism and autoxidation that are mostly based on expert rules since the experimental data available for their support is not sufficiently comprehensive. Toxtree also contains structural alerts for protein binding, and its rules are phrased to partly include pre- or pro-haptens.

An external validation study published in 2013 came to the conclusion that mechanism-based models performed better than statistical models. The study further identified a correlation of predictive accuracy and stringency of the domain definition [78]. TIMES-SS achieved the best performance. This model provided perfect predictions but had the major disadvantage of only covering 16% of the test substances within its reliability domain. Generally, prediction of skin sensitization by in silico models works well for those substances that are sensitizers and contain mechanistic alerts (and do not require enzymatic or non-enzymatic transformations). To date, the sub-categorization of test substances into the GHS Sub-categories 1A and 1B cannot be adequately modelled.

Conclusion

Skin toxicology is one of the first areas of modern toxicology in which mechanistic information, relevant to humans, is systematically generated by in vitro and in silico methods yielding predictivities of human hazard that are sufficiently accurate to allow abandoning the traditional, descriptive animal models. Nevertheless, in vitro and in silico methods are criticized for being too simplistic or incomplete (in the case of skin sensitization for not

covering all four key events of the AOP). Still, several legal provisions prescribe animal models with their own specific imperfections and limitations (such as the LLNA and guinea pig tests). As Box and Draper [79] concluded, *essentially, all models are wrong, but some are useful.* The example of dermal toxicity testing and, specifically, skin sensitization testing underlines that modern toxicological methods have proven to be reliable and useful.

Acknowledgments Ursula G. Sauer (Scientific Consultancy—Animal Welfare) has edited and proofread the manuscript.

References

1. Harber MD, Baer RL. Pathogenic mechanisms of drug-induced photosensitivity. J Invest Dermatol. 1972;58:327–42.
2. Ankley GT, Bennett RS, Erickson RJ, Hoff DJ, Hornung MW, Johnson RD, et al. Adverse outcome pathways: a conceptual framework to support ecotoxicology research and risk assessment. Environ Toxicol Chem. 2010;29:730–41.
3. Russell WMS, Burch RL. The principles of humane experimental technique. Reprinted by UFAW, 1992: 8 Hamilton close, south Mimms, potters bar, Herts EN6 3QD England. London, UK: Methuen; 1959. pp. 238.
4. EP and Council of the EU. Directive 2010/63/EU of the European Parliament and of the Council of 22 September 2010 on the protection of animals used for scientific purposes. O.J. L 276/33, 20 2010.
5. Aardema MJ, Barnett BB, Mun GC, Dahl EL, Curren RD, Hewitt NJ, et al. Evaluation of chemicals requiring metabolic activation in the EpiDerm™ 3D human reconstructed skin micronucleus (RSMN) assay. Mutat Res. 2013;750:40–9.
6. UN. United Nations globally harmonized system of classification and Labelling of chemicals (GHS). 6th revised ed. 2015.
7. EP and Council of the EU. Regulation (EC) no 1272/2008 of the European Parliament and of the Council of 16 December 2008 on classification, labelling and packaging of substances and mixtures, amending and repealing directives 67/548/EEC and 1999/45/EC, and amending regulation (EC) no 1907/2006. O.J L 353/1, 31 2008.
8. EU Commission. Commission regulation (EU) no 605/2014 of 5 June 2014 amending, 6th adaptation to technical and scientific progress, regulation (EC) no 1272/2008 of the European Parliament and of the

Council on classification, labelling and packaging of substances and mixtures. O.J., 6 June 2014. pp. 36–49.

9. EP and Council of the EU. Regulation (EC) no 1907/2006 of the European Parliament and of the Council of 18 December 2006 concerning the registration, evaluation, authorisation and restriction of chemicals (REACH), establishing a European chemicals agency, amending Directive 1999/45/EC and repealing Council regulation (EEC) no 793/93 and Commission regulation (EC) no 1488/94 as well as Council Directive 76/769/EEC and Commission directives 91/155/EEC, 93/67/EEC, 93/105/EC and 2000/21/EC. O.J. L 396/1, 30 2006.

10. OECD. OECD guidance document on the validation of (quantitative) structure-activity relationship [(Q)SAR] models. ENV/JM/MONO(2007)2 OECD, Paris, 30 Mar 2007.

11. Worth A, Barroso J, Bremer S, Burton J, Casati S, Coecke S et al. Alternative methods for regulatory toxicology – a state-of-the-art review. JRC science and policy reports EUR 26797 EN. 2014.

12. Draize JH, Woodard G, Calvery HO. Methods for the study of irritation and toxicity of substances applied topically to the skin and mucous membranes. J Pharmacol Exp Ther. 1944;82:377–90.

13. OECD. Series on testing and assessment no. 137 Explanatory Background Document to the OECD Draft Test Guideline on in vitro Skin Irritation Testing ENV/JM/MONO(2010)36 OECD, Paris, 16 Sep 2010.

14. Basketter DA, York M, McFadden JP, Robinson MK. Determination of skin irritation potential in the human 4-h patch test. Contact Dermatitis. 2004;51:1–4.

15. Jirova D, Basketter D, Liebsch M, Bendova H, Kejlova K, Marriott M, et al. Comparison of human skin irritation patch test data with in vitro skin irritation assays and animal data. Contact Dermatitis. 2010;62:109–16.

16. Robinson MK, McFadden JP, Basketter DA. Validity and ethics of the human 4-h patch test as an alternative method to assess acute skin irritation potential. Contact Dermatitis. 2001;45:1–12.

17. Deshmukh GR, Kumar KH, Reddy PV, Rao BS. In vitro skin corrosion: human skin model test – a validation study. Toxicol In Vitro. 2012;26:1072–4.

18. Hoffmann J, Heisler E, Karpinski S, Losse J, Thomas D, Siefken W, et al. Epidermal-skin-test 1000 (EST-1000) – a new reconstructed epidermis for in vitro skin corrosivity testing. Toxicol In Vitro. 2005;19:925–9.

19. Kandarova H, Hayden P, Klausner M, Kubilus J, Sheasgreen J. An in vitro skin irritation test (SIT) using the EpiDerm reconstructed human epidermal (RHE) model. J Vis Exp. 2009;29, pii: 1366. doi: 10.3791/1366.

20. Kandárová H, Hayden P, Klausner M, Kubilus J, Kearney P, Sheasgreen J. In vitro skin irritation testing: improving the sensitivity of the EpiDerm skin irritation test protocol. Altern Lab Anim. 2009;37:671–89.

21. Kojima H, Ando Y, Idehara K, Katoh M, Kosaka T, Miyaoka E, et al. Validation study of the in vitro skin irritation test with the LabCyte EPI-MODEL24. Altern Lab Anim. 2012;40:33–50.

22. Spielmann H, Hoffmann S, Liebsch M, Botham P, Fentem JH, Eskes C, et al. The ECVAM international validation study on in vitro tests for acute skin irritation: report on the validity of the EPISKIN and EpiDerm assays and on the skin integrity function test. Altern Lab Anim. 2007;5:559–601.

23. Gerner I, Zinke S, Graetschel G, Schlede E. Development of a decision support system for the introduction of alternative methods into local irritancy/corrosivity testing strategies. Creation of fundamental rules for a decision support system. Altern Lab Anim. 2000;28:665–98.

24. Gerner I, Graetschel G, Kahl J, Schlede E. Development of a decision support system for the introduction of alternative methods into local irritation/corrosion testing strategies: development of a relational data base. Altern Lab Anim. 2000;28:11–28.

25. Rorije E, Hulzebos E. Evaluation of (Q)SARs for the prediction of skin irritation/corrosion potential. Physicochemical exclusion rules. Final report for ECB contract IHCP.B430206. European Commission, Joint Research Centre. 2005. https://eurl-ecvam.jrc.ec.europa.eu/laboratories-research/predictive_toxicology/information-sources/qsar-document-area/Evaluation_of_Skin_Irritation_QSARs.pdf

26. Gallegos Saliner A, Tsakovska I, Pavan M, Patlewicz G, Worth AP. Evaluation of SARs for the prediction of skin irritation/corrosion potential. Structural inclusion rules in the BfR decision support system. SAR QSAR Environ Res. 2007;18:331–42.

27. Mombelli E. An evaluation of the predictive ability of the QSAR software packages, DEREK, HAZARDEXPERT and TOPKAT, to describe chemically-induced skin irritation. Altern Lab Anim. 2008;36:15–24.

28. OECD, 2014. OECD guidance document on an integrated approach on testing and assessment (IATA) for skin corrosion and irritation. Series on testing and assessment no. 203, OECD, Paris, Jul 2014.

29. Thong H-Y, Maibach HI. Photosensitivity induced by exogenous agents: phototoxicity and photoallergy. In: Roberts MS, Walters KA, editors. Dermal absorption and toxicity assessment. 2nd ed. New York: Informa Healthcare; 2007, ISBN: 978-0849375910.

30. Epstein JH. Phototoxicity and photoallergy in man. J Am Acad Dermatol. 1983;8:141–7.

31. ICH. International conference on harmonisation of technical requirements for registration of pharmaceuticals for human use. ICH harmonised tripartite guideline: Photosafety evaluation of pharmaceuticals. Current step 4 version, 13 2013.

32. Jirova D, Liebsch M, Basketter D, Spiller E, Kejlova K, Bendova H, et al. Comparison of human skin irritation and photo-irritation patch test data with cellular in vitro assays and animal in vivo data. AATEX. 2007;14(Special Issue):359–65.

33. Kim K, Park H, Lim KM. Phototoxicity: its mechanism and animal alternative test methods. Toxicol Res. 2015;31:97–104.

34. Lasarow RM, Isseroff RR, Gomez EC. Quantitative *in vitro* assessment of phototoxicity by a fibroblast-neutral red assay. J Invest Dermatol. 1992;98:725–9.

35. Ceridono M, Tellner P, Bauer D, Barroso J, Alépée N, Corvi R, et al. The 3T3 neutral red uptake phototoxicity test: practical experience and implications for phototoxicity testing – the report of an ECVAM-EFPIA workshop. Regul Toxicol Pharmacol. 2012;63:480–8.

36. Liebsch M, Traue D, Barrabas C, Spielmann H, Gerberick F, Cruse L, et al. Prevalidation of the Epiderm™ phototoxicity test. In: Clark D, Lisansky S, Macmillan R, editors. Alternatives to animal testing II: Proceedings of 2nd International Science Conference organised by the European Cosmetic Industry, Brussels. Newbury: Belgium CPL Press; 1999. p. 160–6.

37. Onoue S, Hosoi K, Wakuri S, Iwase Y, Yamamoto T, Matsuoka N, et al. Establishment and intra−/interlaboratory validation of a standard protocol of reactive oxygen species assay for chemical photosafety evaluation. J Appl Toxicol. 2013;33:1241–50.

38. OECD. OECD guideline for the testing of chemicals. Draft proposal for a new test guideline ROS (Reactive Oxygen Species) assay for photosafety OECD, Paris, Nov 2016.

39. Haranosono Y, Kurata M, Sakaki H. Establishment of an *in silico* phototoxicity prediction method by combining descriptors related to photo-absorption and photo-reaction. J Toxicol Sci. 2014;39:655–64.

40. Ringeissen S, Marrot L, Note R, Labarussiat A, Imbert S, Todorov M, et al. Development of a mechanistic SAR model for the detection of phototoxic chemicals and use in an integrated testing strategy. Toxicol In Vitro. 2011;25:324–34.

41. OECD. Series on testing and assessment, no. 168. The adverse outcome pathway for skin sensitisation initiated by covalent binding to proteins. Part 1: scientific evidence. ENV/JM/MONO(2012)10/PART 1. OECD, Paris, 4 May 2012.

42. OECD. Series on testing and assessment no. 168. The adverse outcome pathway for skin sensitisation initiated by covalent binding to proteins. Part 2: use of the AOP to develop chemical categories and integrated assessment and testing approaches. ENV/JM/MONO(2012)10/PART 2. OECD, Paris, 4 May 2012.

43. Gerberick F, Aleksic M, Basketter D, Casati C, Karlberg AT, Kern P, et al. Chemical reactivity measurement and the predicitve identification of skin sensitisers. Altern Lab Anim. 2008;36:215–42.

44. Matzinger P. The danger model: a renewed sense of self. Science. 2002;296:301–5.

45. McFadden JP, Basketter DA. Contact allergy, irritancy and 'danger'. Contact Dermatitis. 2000;42:123–7.

46. Kimber I, Cumberbatch M. Dendritic cells and cutaneous immune responses to chemical allergens. Toxicol Appl Pharmacol. 1992;117:137–46.

47. Ryan CA, Kimber I, Basketter DA, Pallardy M, Gildea LA, Gerberick GF. Dendritic cells and skin sensitization: biological roles and uses in hazard identification. Toxicol Appl Pharmacol. 2007;221:384–94.

48. Steinman RM. The dendritic cell system and its role in immunogenicity. Annu Rev Immunol. 1991;9:271–96.

49. Banchereau J, Steinman RM. Dendritic cells and the control of immunity. Nature. 1998;392:245–52.

50. Ahn I, Kim TS, Jung ES, Yi JS, Jang WH, Jung KM, et al. Performance standard-based validation study for local lymph node assay: 5-bromo-2-deoxyuridine-flow cytometry method. Regul Toxicol Pharmacol. 2016;80:183–94.

51. Basketter D, Kolle SN, Schrage A, Honarvar N, Gamer AO, van Ravenzwaay B, et al. Experience with local lymph node assay performance standards using standard radioactivity and nonradioactive cell count measurements. J Appl Toxicol. 2012;32:590–6.

52. Ball N, Cagen S, Carrillo JC, Certa H, Eigler D, Emter R, et al. Evaluating the sensitization potential of surfactants: integrating data from the local lymph node assay, guinea pig maximization test, and *in vitro* methods in a weight-of-evidence approach. Regul Toxicol Pharmacol. 2011;60:389–400.

53. Basketter D, Ball N, Cagen S, Carrillo JC, Certa H, Eigler D, et al. Application of a weight of evidence approach to assessing discordant sensitisation datasets: implications for REACH. Regul Toxicol Pharmacol. 2009;55:90–6.

54. Kreiling R, Hollnagel HM, Hareng L, Eigler D, Lee MS, Griem P, et al. Comparison of the skin sensitizing potential of unsaturated compounds as assessed by the murine local lymph node assay (LLNA) and the guinea pig maximization test (GPMT). Food Chem Toxicol. 2008;46:1896–904.

55. Hoffmann S. LLNA variability: an essential ingredient for a comprehensive assessment of non-animal skin sensitization test methods and strategies. ALTEX. 2015;32:379–83.

56. Luechtefeld T, Maertens A, Russo DP, Rovida C, Zhu H, Hartung T. Analysis of publically available skin sensitization data from REACH registrations 2008–2014. ALTEX. 2016;33:135–48.

57. Kolle SN, Basketter DA, Casati S, Stokes WS, Strickland J, van RB, et al. Performance standards and alternative assays: practical insights from skin sensitization. Regul Toxicol Pharmacol. 2013;65:278–85.

58. Leontaridou M, Urbisch D, Kolle SN, Ott K, Mulliner DS, Gabbert S, Landsiedel R. Quantification of the borderline range and implications for evaluating non-animal testing methods' precision. ALTEX. 2017; doi:10.14573/altex.1606271.

59. Mehling A, Eriksson T, Eltze T, Kolle S, Ramirez T, Teubner W, et al. Non-animal test methods for predicting skin sensitization potentials. Arch Toxicol. 2012;86:1273–95.

60. Andreas N, Caroline B, Leslie F, Frank G, Kimberly N, Allison H, et al. The intra- and inter-laboratory reproducibility and predictivity of the KeratinoSens assay to predict skin sensitizers *in vitro*: results of a ring-study in five laboratories. Toxicol In Vitro. 2011;25(3):733–44.

61. Ramirez T, Mehling A, Kolle SN, Wruck CJ, Teubner W, Eltze T, et al. LuSens: a keratinocyte based ARE reporter gene assay for use in integrated testing strategies for skin sensitization hazard identification. Toxicol In Vitro. 2014;28:1482–97.

62. Ramirez T, Stein N, Aumann A, Remus T, Edwards A, Norman KG, et al. Intra- and inter-laboratory reproducibility and accuracy of the LuSens assay: a reporter gene-cell line to detect keratinocyte activation by skin sensitizers. Toxicol In Vitro. 2016;32:278–86.

63. Kobayashi A, Ohta T, Yamamoto M. Unique function of the Nrf2-Keap1 pathway in the inducible expression of antioxidant and detoxifying enzymes. Methods Enzymol. 2004;378:273–86.

64. Ashikaga T, Yoshida Y, Hirota M, Yoneyama K, Itagaki H, Sakaguchi H, et al. Development of an *in vitro* skin sensitization test using human cell lines: the human cell line activation test (h-CLAT). I. Optimization of the h-CLAT protocol. Toxicol In Vitro. 2006;20:767–73.

65. Ashikaga T, Sakaguchi H, Sono S, Kosaka N, Ishikawa M, Nukada Y, et al. A comparative evaluation of *in vitro* skin sensitization tests: the human cell-line activation test (h-CLAT) versus the local lymph node assay (LLNA). Altern Lab Anim. 2010;38:275–84.

66. Sauer UG, et al. Local tolerance testing under REACH: accepted non-animal methods are not on equal footing with animal tests. ATLA. 2016;443:281.

67. Rovida C, Alépée N, Api AM, Basketter DA, Bois FY, Caloni F, et al. Integrated testing strategies (ITS) for safety assessment. ALTEX. 2015;32:25–40.

68. OECD. OECD guidance document on the reporting of defined approaches and individual information sources to be used within integrated approaches to testing and assessment (IATA) for skin sensitization ENV/JM/MONO(2016)29 (Ann. I: case studies; Ann. 2: information sources used within the case studies). 2016. http://www.Oecd.Org/officialdocuments/publicdisplay documentpdf/?Cote=env/jm/mono(2016)29&doclang uage=en

69. Jaworska J, Dancik Y, Kern P, Gerberick F, Natsch A. Bayesian integrated testing strategy to assess skin sensitization potency: from theory to practice. J Appl Toxicol. 2013;33:1353–64.

70. Tsujita-Inoue K, Hirota M, Ashikaga T, Atobe T, Kouzuki H, Aiba S. Skin sensitization risk assessment model using artificial neural network analysis of data from multiple *in vitro* assays. Toxicol In Vitro. 2014;28:626–39.

71. van der Veen JW, Rorije E, Emter R, Natsch A, van Loveren H, Ezendam J. Evaluating the performance of integrated approaches for hazard identification of skin sensitizing chemicals. Regul Toxicol Pharmacol. 2014;69:371–9.

72. Bauch C, Kolle SN, Ramirez T, Eltze T, Fabian E, Mehling A, et al. Putting the parts together: combining *in vitro* methods to test for skin sensitizing potentials. Regul Toxicol Pharmacol. 2012;63:489–504.

73. Natsch A, Ryan CA, Foertsch L, Emter R, Jaworska J, Gerberick F, et al. A dataset on 145 chemicals tested in alternative assays for skin sensitization undergoing prevalidation. J Appl Toxicol. 2013;33:1337–52.

74. Urbisch D, Mehling A, Guth K, Ramirez T, Honarvar N, Kolle SN, et al. Assessing skin sensitization hazard in mice and men using non-animal test methods. Regul Toxicol Pharmacol. 2015;71:337–51.

75. Aptula AO, Patlewicz G, Roberts DW, Schultz TW. Non-enzymatic glutathione reactivity and *in vitro* toxicity: a non-animal approach to skin sensitization. Toxicol In Vitro. 2006;20:239–47.

76. Urbisch D, Becker M, Honarvar N, Kolle SN, Mehling A, Teubner W, Wareing B, Landsiedel R. Assessment of pre- and pro-haptens using non-animal test methods for skin sensitization. Chem Res Toxicol. 2016;29:901–13.

77. Patlewicz G, Worth A. Review of data sources, QSARs and integrated testing strategies for skin sensitisation. EUR 23225 EN. European Commission, Joint Research Centre. JRC Scientific and Technical Reports. 2008.

78. Teubner W, Mehling A, Schuster PX, Guth K, Worth A, Burton J, et al. Computer models versus reality: how well do in *silico* models currently predict the sensitization potential of a substance. Regul Toxicol Pharmacol. 2013;67:468–85.

79. Box GEP, Draper NR. Empirical model building and response surfaces. New York, NY: Wiley; 1987. p. 424.

Part II

Environmental Threats to the Skin

Contact Allergy

5

Stefan F. Martin

Abbreviations

ACD	Allergic contact dermatitis
CHS	Contact hypersensitivity
DAMPs	Damage-associated molecular patterns
DC	Dendritic cell
DNFB	2,4-Dinitrofluorobenzene
ICD	Irritant contact dermatitis
NLR	NOD-like receptor
PAMPs	Pathogen-associated molecular patterns
PRR	Pattern recognition receptor
ROS	Reactive oxygen species
TLR	Toll-like receptor
TNCB	2,4,6-Trinitrochlorobenzene

5.1 Contact Dermatitis

Contact allergens are small organic chemicals or metal ions. As such they are too small to be recognized by the immune system. Therefore, they are also called haptens (half-antigens) and are obligatory reactive (electrophilic or complex forming). Chemical reactivity is the main feature that differ-

entiates contact allergens from irritants. The latter are, for example, detergents such as SDS that cause irritant contact dermatitis (ICD) but fail to activate an adaptive immune response. They exert toxic effects on skin cells but also seem to evoke some innate immune responses. Contact allergens cause allergic contact dermatitis (ACD). The prevalence is high with 15–20% of the general population sensitized to at least one contact allergen and 5–10% developing clinically manifest disease at least once a year [1, 2]. Occupational ICD and ACD are some of the most important occupational diseases [3, 4]. Many chemicals are encountered in the workplace and can cause severe problems [5]. Given the prevalence of contact dermatitis and its impact on human health, the urgent need for the development of novel mechanism-based targeted therapies is obvious.

Sensitization to a contact allergen can occur upon first skin contact and involves the induction of an innate inflammatory immune response and eventually a contact allergen-specific T cell response [6, 7]. Elicitation of ACD requires repeated skin contact with the same contact allergen and results in the recruitment of circulating contact allergen-specific T cells into the skin. The T cells are then activated when they recognize the contact allergen in the context of MHC molecules on skin cells. They promote inflammation due to their cytotoxic effector function and secretion of cytokines such as IFN-γ and IL-17 [8]. The immune response is then rapidly downregulated

S.F. Martin, Ph.D.
Allergy Research Group, Department
of Dermatology, Medical Center – University of
Freiburg, Hauptstrasse 7, 79104 Freiburg, Germany

Faculty of Medicine, University of Freiburg,
Hauptstrasse 7, 79104 Freiburg, Germany
e-mail: stefan.martin@uniklinik-freiburg.de

© Springer International Publishing Switzerland 2018
J. Krutmann, H.F. Merk (eds.), *Environment and Skin*,
https://doi.org/10.1007/978-3-319-43102-4_5

by regulatory T cells and other cells with immuno-regulatory function such as NKT cells [7, 9].

5.2 Diagnosis of ACD

Productive sensitization is clinically inapparent but can be diagnosed by the patch test [2]. In this in vivo test, test substances are applied to the back skin for 48 hours. If sensitization to a test chemical exists, a red eczematous skin reaction is seen in the specific test area. This ACD is caused by contact allergen-specific T cells of the sensitized individual that are recruited to the inflamed skin and exert their effector function such as cytotoxic activity against cells which present contact allergen in the context of MHC molecules. The patch test is a very old method that requires reading by trained and experienced experts. One of the future goals is the identification of, ideally, circulating biomarkers that unequivocally identify ACD and allow its distinction from ICD and from other forms of eczema. Recent studies are pointing in this direction. Here, global profiling technologies were used to identify contact allergen-specific or eczema subtype-specific gene signatures using human skin biopsies. The comparison of different contact allergens revealed commonly regulated genes but also genes that are specific for individual allergens [10]. The intraindividual comparison of patients with psoriatic, atopic, or nickel-induced eczema also revealed gene signatures that are common or specific for the eczema subtype [11]. The identification of allergen- and disease-specific biomarkers will pave the way to modern, molecular diagnostics. Moreover, such studies will reveal novel potential drug targets for causative therapies.

5.3 Chemical Reactivity and Immune System Activation

Contact allergens are reactive low molecular weight organic chemicals or metal ions. Their reaction with biomolecules either by covalent binding or by complex formation is essential for their antigenicity and immunogenicity. However, contact allergens are very special due to their ability to simultaneously activate the innate immune system and to form T cell epitopes.

The formation of T cell epitopes requires protein reactivity of contact allergens. They may directly bind to peptides presented by MHC molecules on the cell surface or, as is the case for metal ions, form complexes with peptides and MHC molecules or MHC molecules alone in a peptide-independent manner. In addition, extra- and intracellular proteins can be chemically modified which are then processed to generate contact allergen-modified peptides that are displayed on MHC molecules on the surface of antigen-presenting cells (APC) [8].

Another, essential consequence of the chemical reactivity of contact allergens is the activation of the innate immune system, a prerequisite for the activation of the adaptive immune system [12, 13]. Here, contact allergens are very peculiar since they engage pathways characteristic for innate immune responses to infections, and this can happen under sterile conditions. This special type of inflammation induced by xenobiotic substances was called xenoinflammation to distinguish it from microbial inflammation or auto-inflammation [12, 14].

5.4 The Innate Molecular Immune Response to Contact Allergens

The surprising result of studies addressing the innate immune system activation by (the few analyzed) contact allergens was that they trigger the same signaling pathways that are triggered by infectious agents. During an infection some pathogens activate pattern recognition receptors (PRRs) such as the membrane-associated Toll-like receptors (TLRs) and the cytosolic NOD-like receptor (NLR) NLRP3, a component of the caspase-1 activating NLRP3 inflammasome. These receptors recognize components of bacteria and viruses such as DNA or RNA, bacterial cell wall components or bacterial toxins commonly designated pathogen-associated molecular patterns (PAMPs) and trigger the production of pro-inflammatory cytokines and chemokines. Vaccination against protein antigens usually requires addition of such PAMPs, commonly known as adjuvants. Infectious agents also trigger the production of pro-inflammatory reactive oxygen species (ROS) which promote TLR and NLRP3 inflammasome activation. As mentioned above, contact allergens are very peculiar: they can activate both the innate and the adaptive immune system, i.e., they possess auto-adjuvant activity. In recent years, significant progress has been made in the elucidation of the mechanistic basis for this special feature of contact allergens. Contact allergens efficiently activate PRRs. For the TLRs direct and indirect activation has been shown. The metal ions nickel and cobalt directly bind to conserved histidine residues in the human LPS receptor TLR4 causing its dimerization and signaling in the absence of the cognate ligand LPS [15, 16]. These histidine residues are missing in the mouse TLR4, which explains the failure of nickel to induce ACD in the mouse contact hypersensitivity (CHS) model unless an adjuvant such as LPS is added. Organic chemicals such as oxazolone and 2,4,6-trinitrochlorobenzene (TNCB) activate TLR2 and TLR4 indirectly as shown in the CHS model [17]. They cause the degradation of the extracellular matrix component hyaluronic acid (HA). HA fragments can then activate these TLRs. In addition they cause oxidative stress and activation of the antioxidant phase II response due to their induction of reactive oxygen species (ROS) [18, 19]. All of these contact allergens activate the NLRP3 inflammasome. In the case of the organic chemicals, it was shown that this is mediated by causing cellular stress resulting in the release of ATP and, consequently, the triggering of inflammasome activation via the ATP receptor P2X7R [20].

Another signaling pathway triggered by contact allergens involves unknown receptors that couple to the kinase Syk [21]. It was shown that the production of IL-1β was mediated by signaling via the kinase Syk coupling to the adaptor CARD9 and Bcl10 for NF-κB activation. The processing of immature pro-IL-1β requires cas-pase-1 activation via the NLRP3 inflammasome. This was dependent on ROS production but independent of CARD9/Bcl10 or the TLR/IL-1R-associated adaptor protein MyD88. This study implies that contact allergens directly or indirectly engage an as yet unknown immunoreceptor tyrosine-based activation motif (ITAM) containing receptor that signals via Syk.

5.5 The Innate Cellular Immune Response to Contact Allergens

The triggering of the innate immune response by contact allergens does not only involve the activation of hematopoietic cells. Structural skin cells such as keratinocytes and dermal fibroblasts participate in the immune response. They express PRRs like the hematopoietic cells. The interplay of many different skin resident and migratory cell types orchestrates the innate inflammatory response in a highly dynamic process. One example of such an interplay is the cross-talk between mast cells, neutrophils, and DCs in the sensitization phase of ACD. The use of genetically engineered mouse strains which lack mast cells or allow for mast cell depletion has revealed their important pro-inflammatory role in CHS [22]. Depletion before sensitization or lack of mast cells strongly reduced ear swelling responses to DNFB or FITC. It was then shown that neutrophils which infiltrate the skin in CHS also have an important pro-inflammatory role [23]. In the absence of mast cells, their skin infiltration was compromised. Moreover, DCs failed to efficiently migrate to skin-draining lymph nodes, and T cell priming was abrogated. In addition, neutrophils were also required in the elicitation phase of CHS in order to enable T cell recruitment to the inflamed skin. It remains to be shown how mast cells and neutrophils are activated by contact allergens. Also here PRRs may play a role as recently demonstrated for TLR-dependent neutrophil activation in graft-versus-host disease [24]. Interestingly, a recent study revealed that neutrophils leave chemokine cues for CD8+ T cell migration in the airways. In an influenza infection model, the authors demonstrated CXCR4-dependent CD8 T cell migration along trails of packed CXCL12 [25]. Moreover, the effector function of the T cells was impaired in the absence of neutrophils as was proliferation in another study demonstrating neutrophil migration to lymph nodes in a bacterial skin infection model [26]. Thus infiltrated neutrophils may help to guide T cells into inflamed tissue sites and modulate T cell function.

5.6 Identification of Contact Allergens by In Vitro Assays: Alternatives to Animal Testing

Key events in the skin sensitization by chemicals have been summarized in the adverse outcome pathway (AOP) for skin sensitization [27]. Replacement of animal testing for the assessment of the skin sensitizing potential of chemicals has resulted in the development of in vitro assays which cover the different steps of the AOP. The outcome of these efforts will be the development of an integrated testing strategy that combines several assays [28–31]. Up to now three in vitro assays have been fully validated and are now OECD guideline tests. The direct peptide reactivity assay (DPRA, OECD guideline test 442C) detects chemical reactivity of test substances based on the depletion of model peptides that contain lysine or cysteine residues. Covalent binding of contact allergens causes a mass shift and disappearance of the mass peak of the unmodified peptide. The second assay is the ARE-Nrf2 Luciferase Test Method (OECD guideline test 442D) as represented by the KeratinoSens Assay. This assay is based on the activation of the antioxidant phase 2 response [32, 33] and detects the activation of the transcription factor Nrf2 in a luciferase reporter system in the human keratinocyte cell line HaCaT. A third test is the human cell line activation test (hCLAT, OECD guideline test 442E)). In this assay human monocytic leukemia THP-1 cells are stimulated by test chemicals, and the upregulation of the costimulatory molecule CD86 is detected by flow cytometry.

Ideally, an integrated approach to testing and assessment (IATA) should combine a minimal number of assays, but currently, it is not clear yet which combination of assays is optimal for reliable hazard identification.

5.7 Heterologous Innate Immune Stimulation

The hazard and risk assessment in the mouse local lymph node assay (OECD TG 429), the gold standard for contact allergen identification, as well as the current hazard identification in in vitro assays is performed with single substances. However, final products are usually mixtures and formulations containing a number of different substances, among them several contact allergens, detergents, and preservatives. It has to be considered that such combinations may facilitate chemical penetration into the skin, thereby increasing local concentrations. In addition, the combination of contact allergens and irritants may lead to synergistic or additive effects that amplify innate immune and stress responses [34–36]. This may lead to sensitizations not observed with single compounds which can be explained mechanistically [37]. A good example is the TLR system. Of the 10 human and 13 mouse TLRs most signal via the adaptor protein MyD88 to activate NF-κB. Therefore, simultaneous direct or indirect triggering of different TLRs by contact allergens may result in the amplification of inflammation. This may explain additive or synergisitc effects as seen with mixtures and formulations. It is also possible to substitute for a missing innate immune stimulus not given by weak contact allergens by providing a heterologous stimulus by other contact allergens or even by infection. In our CHS model, we have previously shown that mice lacking both TLR2 and TLR4 or TLR4 and IL-12Rβ2 are resistant to CHS. However, triggering of TLR9 on DCs or injection of synthetic TLR9 ligands into the skin of mice prior to sensitization with TNCB restored CHS [17].

These findings show the importance of the concept of heterologous innate immune stimulation [37]. A simplified view is that the DC does not matter by which stimuli it is activated. As long as proper DC activation and polarization occurs to allow for the priming of contact allergen-specific T cells, it does not matter whether the contact allergen that is recognized by the T cells has provided autologous innate immune stimulation at all or sufficiently strong. Heterologous innate immune stimuli as given by other contact allergens or ingredients of formulations and mixtures or even by infections can provide additive or synergistic effects or even fully substitute for missing autologous stimuli.

Conclusions

ACD and ICD are important inflammatory skin diseases with great socioeconomic impact. The elucidation of the molecular and cellular pathomechanisms is moving forward and provides essential mechanistic insights. This is the basis for mechanism-based toxicological research and in vitro assay development to replace animal testing for contact allergen identification and for the identification of novel drug targets and the resulting design of targeted causative therapies. Biomarker discovery will improve diagnosis and promote our mechanistic understanding of chemical-induced skin disease.

References

1. Peiser M, Tralau T, Heidler J, Api AM, Arts JH, Basketter DA, English J, Diepgen TL, Fuhlbrigge RC, Gaspari AA, Johansen JD, Karlberg AT, Kimber I, Lepoittevin JP, Liebsch M, Maibach HI, Martin SF, Merk HF, Platzek T, Rustemeyer T, Schnuch A, Vandebriel RJ, White IR, Luch A. Allergic contact dermatitis: epidemiology, molecular mechanisms, in vitro methods and regulatory aspects. Current knowledge assembled at an international workshop at BfR, Germany. Cell Mol Life Sci. 2012;69(5):763–81. doi:10.1007/s00018-011-0846-8.
2. Brasch J, Becker D, Aberer W, Bircher A, Kranke B, Jung K, Przybilla B, Biedermann T, Werfel T, John SM, Elsner P, Diepgen T, Trautmann A, Merk HF, Fuchs T, Schnuch A. Guideline contact dermatitis: S1-Guidelines of the German Contact Allergy Group (DKG) of the German Dermatology Society (DDG), the Information Network of Dermatological Clinics (IVDK), the German Society for Allergology and Clinical Immunology (DGAKI), the Working Group for Occupational and Environmental Dermatology (ABD) of the DDG, the Medical Association of German Allergologists (AeDA), the Professional Association of German Dermatologists (BVDD) and the DDG. Allergo J Int. 2014;23(4):126–38. doi:10.1007/s40629-014-0013-5.
3. Holness DL. Occupational skin allergies: testing and treatment (the case of occupational allergic contact dermatitis). Curr Allergy Asthma Rep. 2014;14(2):410. doi:10.1007/s11882-013-0410-8.
4. Wiszniewska M, Walusiak-Skorupa J. Recent trends in occupational contact dermatitis. Curr Allergy Asthma Rep. 2015;15(7):43. doi:10.1007/s11882-015-0543-z.
5. Fyhrquist N, Lehto E, Lauerma A. New findings in allergic contact dermatitis. Curr Opin Allergy Clin Immunol. 2014;14(5):430–5. doi:10.1097/ACI.0000000000000092.
6. Martin SF. Immunological mechanisms in allergic contact dermatitis. Curr Opin Allergy Clin Immunol. 2015;15(2):124–30. doi:10.1097/ACI.0000000000000142.
7. Vocanson M, Hennino A, Rozieres A, Poyet G, Nicolas JF. Effector and regulatory mechanisms in allergic contact dermatitis. Allergy. 2009;64(12):1699–714. doi:10.1111/j.1398-9995.2009.02082.x.
8. Martin SF, Esser PR, Schmucker S, Dietz L, Naisbitt DJ, Park BK, Vocanson M, Nicolas JF, Keller M, Pichler WJ, Peiser M, Luch A, Wanner R, Maggi E, Cavani A, Rustemeyer T, Richter A, Thierse HJ, Sallusto F. T-cell recognition of chemicals, protein allergens and drugs: towards the development of in vitro assays. Cell Mol Life Sci. 2010;67(24):4171–84. doi:10.1007/s00018-010-0495-3.
9. Goubier A, Vocanson M, Macari C, Poyet G, Herbelin A, Nicolas JF, Dubois B, Kaiserlian D. Invariant NKT cells suppress CD8(+) T-cell-mediated allergic contact dermatitis independently of regulatory CD4(+) T cells. J Invest Dermatol. 2013;133(4):980–7. doi:10.1038/jid.2012.404.
10. Dhingra N, Shemer A, Correa da Rosa J, Rozenblit M, Fuentes-Duculan J, Gittler JK, Finney R, Czarnowicki T, Zheng X, Xu H, Estrada YD, Cardinale I, Suarez-Farinas M, Krueger JG, Guttman-Yassky E. Molecular profiling of contact dermatitis skin identifies allergen-dependent differences in immune response. J Allergy Clin Immunol. 2014;134(2):362–72. doi:10.1016/j.jaci.2014.03.009.
11. Quaranta M, Knapp B, Garzorz N, Mattii M, Pullabhatla V, Pennino D, Andres C, Traidl-Hoffmann C, Cavani A, Theis FJ, Ring J, Schmidt-Weber CB, Eyerich S, Eyerich K. Intraindividual genome expression analysis reveals a specific molecular signature of psoriasis and eczema. Sci Transl Med. 2014;6(244):244ra290. doi:10.1126/scitranslmed.3008946.

12. Martin SF. Allergic contact dermatitis: xenoinflamma-
    tion of the skin. Curr Opin Immunol. 2012;24(6):720–
    9. doi:10.1016/j.coi.2012.08.003.
13. Kaplan DH, Igyarto BZ, Gaspari AA. Early immune
    events in the induction of allergic contact dermatitis.
    Nat Rev Immunol. 2012;12(2):114–24. doi:10.1038/
    nri3150.
14. Honda T, Egawa G, Grabbe S, Kabashima K. Update
    of immune events in the murine contact hypersensitiv-
    ity model: toward the understanding of allergic contact
    dermatitis. J Invest Dermatol. 2013;133(2):303–15.
    doi:10.1038/jid.2012.284.
15. Schmidt M, Raghavan B, Muller V, Vogl T, Fejer G,
    Tchaptchet S, Keck S, Kalis C, Nielsen PJ, Galanos
    C, Roth J, Skerra A, Martin SF, Freudenberg MA,
    Goebeler M. Crucial role for human Toll-like receptor
    4 in the development of contact allergy to nickel. Nat
    Immunol. 2010;11(9):814–9. doi:10.1038/ni.1919.
16. Raghavan B, Martin SF, Esser PR, Goebeler M,
    Schmidt M. Metal allergens nickel and cobalt facili-
    tate TLR4 homodimerization independently of MD2.
    EMBO Rep. 2012;13(12):1109–15. doi:10.1038/
    embor.2012.155.
17. Martin SF, Dudda JC, Bachtanian E, Lembo A,
    Liller S, Durr C, Heimesaat MM, Bereswill S, Fejer
    G, Vassileva R, Jakob T, Freudenberg N, Termeer
    CC, Johner C, Galanos C, Freudenberg MA. Toll-
    like receptor and IL-12 signaling control suscep-
    tibility to contact hypersensitivity. J Exp Med.
    2008;205(9):2151–62. doi:10.1084/jem.20070509.
18. Esser PR, Wolfle U, Durr C, von Loewenich FD,
    Schempp CM, Freudenberg MA, Jakob T, Martin
    SF. Contact sensitizers induce skin inflammation via
    ROS production and hyaluronic acid degradation.
    PLoS One. 2012;7(7):e41340. doi:10.1371/journal.
    pone.0041340.
19. El Ali Z, Gerbeix C, Hemon P, Esser PR, Martin SF,
    Pallardy M, Kerdine-Romer S. Allergic skin inflam-
    mation induced by chemical sensitizers is con-
    trolled by the transcription factor Nrf2. Toxicol Sci.
    2013;134(1):39–48. doi:10.1093/toxsci/kft084.
20. Weber FC, Esser PR, Muller T, Ganesan J, Pellegatti
    P, Simon MM, Zeiser R, Idzko M, Jakob T, Martin
    SF. Lack of the purinergic receptor P2X(7) results
    in resistance to contact hypersensitivity. J Exp Med.
    2010;207(12):2609–19. doi:10.1084/jem.20092489.
21. Yasukawa S, Miyazaki Y, Yoshii C, Nakaya M, Ozaki
    N, Toda S, Kuroda E, Ishibashi K, Yasuda T, Natsuaki
    Y, Mi-ichi F, Iizasa E, Nakahara T, Yamazaki M,
    Kabashima K, Iwakura Y, Takai T, Saito T, Kurosaki
    T, Malissen B, Ohno N, Furue M, Yoshida H, Hara
    H. An ITAM-Syk-CARD9 signalling axis triggers
    contact hypersensitivity by stimulating IL-1 produc-
    tion in dendritic cells. Nature Commun. 2014;5:3755.
    doi:10.1038/ncomms4755.
22. Dudeck A, Dudeck J, Scholten J, Petzold A,
    Surianarayanan S, Kohler A, Peschke K, Vohringer
    D, Waskow C, Krieg T, Muller W, Waisman A,
    Hartmann K, Gunzer M, Roers A. Mast cells are key
    promoters of contact allergy that mediate the adjuvant
    effects of haptens. Immunity. 2011;34(6):973–84.
    doi:10.1016/j.immuni.2011.03.028.
23. Weber FC, Nemeth T, Csepregi JZ, Dudeck A,
    Roers A, Ozsvari B, Oswald E, Puskas LG, Jakob T,
    Mocsai A, Martin SF. Neutrophils are required for
    both the sensitization and elicitation phase of con-
    tact hypersensitivity. J Exp Med. 2015;212(1):15–22.
    doi:10.1084/jem.20130062.
24. Schwab L, Goroncy L, Palaniyandi S, Gautam S,
    Triantafyllopoulou A, Mocsai A, Reichardt W,
    Karlsson FJ, Radhakrishnan SV, Hanke K, Schmitt-
    Graeff A, Freudenberg M, von Loewenich FD, Wolf
    P, Leonhardt F, Baxan N, Pfeifer D, Schmah O,
    Schonle A, Martin SF, Mertelsmann R, Duyster J,
    Finke J, Prinz M, Henneke P, Hacker H, Hildebrandt
    GC, Hacker G, Zeiser R. Neutrophil granulocytes
    recruited upon translocation of intestinal bacteria
    enhance graft-versus-host disease via tissue dam-
    age. Nat Med. 2014;20(6):648–54. doi:10.1038/
    nm.3517.
25. Lim K, Hyun YM, Lambert-Emo K, Capece T, Bae S,
    Miller R, Topham DJ, Kim M. Neutrophil trails guide
    influenza-specific CD8(+) T cells in the airways.
    Science. 2015;349(6252):aaa4352. doi:10.1126/sci-
    ence.aaa4352.
26. Hampton HR, Bailey J, Tomura M, Brink R, Chtanova
    T. Microbe-dependent lymphatic migration of neutro-
    phils modulates lymphocyte proliferation in lymph
    nodes. Nat Commun. 2015;6:7139. doi:10.1038/
    ncomms8139.
27. MacKay C, Davies M, Summerfield V, Maxwell
    G. From pathways to people: applying the adverse
    outcome pathway (AOP) for skin sensitization to risk
    assessment. ALTEX. 2013;30(4):473–86.
28. Leist M, Hasiwa N, Rovida C, Daneshian M, Basketter
    D, Kimber I, Clewell H, Gocht T, Goldberg A, Busquet
    F, Rossi AM, Schwarz M, Stephens M, Taalman R,
    Knudsen TB, McKim J, Harris G, Pamies D, Hartung
    T. Consensus report on the future of animal-free sys-
    temic toxicity testing. ALTEX. 2014;31(3):341–56.
    doi:10.14573/altex.1406091.
29. Wong CL, Ghassabian S, Smith MT, Lam AL. In vitro
    methods for hazard assessment of industrial chemi-
    cals – opportunities and challenges. Front Pharmacol.
    2015;6:94. doi:10.3389/fphar.2015.00094.
30. Reisinger K, Hoffmann S, Alepee N, Ashikaga T,
    Barroso J, Elcombe C, Gellatly N, Galbiati V, Gibbs
    S, Groux H, Hibatallah J, Keller D, Kern P, Klaric M,
    Kolle S, Kuehnl J, Lambrechts N, Lindstedt M, Millet
    M, Martinozzi-Teissier S, Natsch A, Petersohn D, Pike
    I, Sakaguchi H, Schepky A, Tailhardat M, Templier
    M, van Vliet E, Maxwell G. Systematic evaluation of
    non-animal test methods for skin sensitisation safety
    assessment. Toxicol In Vitro. 2015;29(1):259–70.
    doi:10.1016/j.tiv.2014.10.018.
31. Urbisch D, Mehling A, Guth K, Ramirez T, Honarvar
    N, Kolle S, Landsiedel R, Jaworska J, Kern PS,
    Gerberick F, Natsch A, Emter R, Ashikaga T,
    Miyazawa M, Sakaguchi H. Assessing skin sensitiza-
    tion hazard in mice and men using non-animal test

methods. Regul Toxicol Pharmacol. 2015;71(2):337–51. doi:10.1016/j.yrtph.2014.12.008.

32. Hayes JD, Dinkova-Kostova AT. The Nrf2 regulatory network provides an interface between redox and intermediary metabolism. Trends Biochem Sci. 2014;39(4):199–218. doi:10.1016/j.tibs.2014.02.002.

33. Tebay LE, Robertson H, Durant ST, Vitale SR, Penning TM, Dinkova-Kostova AT, Hayes JD. Mechanisms of activation of the transcription factor Nrf2 by redox stressors, nutrient cues, and energy status and the pathways through which it attenuates degenerative disease. Free Radic Biol Med. 2015;88(Pt B):108–46. doi:10.1016/j.freeradbiomed.2015.06.021.

34. Agner T, Johansen JD, Overgaard L, Volund A, Basketter D, Menne T. Combined effects of irritants and allergens. Synergistic effects of nickel and sodium lauryl sulfate in nickel- sensitized individuals. Contact Dermatitis. 2002;47(1):21–6.

35. Pedersen LK, Johansen JD, Held E, Agner T. Augmentation of skin response by exposure to a combination of allergens and irritants – a review. Contact Dermatitis. 2004;50(5):265–73. doi:10.1111/j.0105-1873.2004.00342.x.

36. Bonefeld CM, Nielsen MM, Rubin IM, Vennegaard MT, Dabelsteen S, Gimenez-Arnau E, Lepoittevin JP, Geisler C, Johansen JD. Enhanced sensitization and elicitation responses caused by mixtures of common fragrance allergens. Contact Dermatitis. 2011;65(6):336–42. doi:10.1111/j.1600-0536.2011.01945.x.

37. Martin SF. Adaptation in the innate immune system and heterologous innate immunity. Cell Mol Life Sci. 2014;71(21):4115–30. doi:10.1007/s00018-014-1676-2.

Contact Urticaria and Contact Urticaria Syndrome

6

Hans F. Merk

Contact urticaria, contact urticaria syndrome (CUS), protein contact dermatitis (PCD), and oral allergy syndrome (OAS) are diseases which are precipitated by immediate contact skin and mucous membrane reactions. They may be the result of a specific immune reaction after sensitization to a high or low molecular weight compound (ICoU), or they are mediated by a nonimmunological reaction which needs no prior sensitization (NICoU). Compounds in numerous environmental materials which can come in contact with the skin are able to induce contact urticaria such as animal dander, plants, foods and flavorings, enzymes, cosmetics including fragrances and preservatives, medications, disinfectants, and metals [1, 2].

Clinically contact urticaria is characterized by a typical triple response—wheal, flare, and itching—of the skin at the site of contact to the eliciting agent. It appears within 30 min after the contact and disappears completely within hours without residual signs and symptoms. The triple response precipitates alone or together with an immediately induced eczema, which can also result, if the exposure to the antigen persists.

Most compounds which induce these reactions belong to cosmetics, plants, vegetables, and food. Contact urticaria has been first described by Fisher and has been recognized as a syndrome by Johnson and Maibach subsequently [3].

Small molecular weight as well as high molecular weight compounds can induce a contact urticaria. In particular, nonimmunological urticaria (NICoU) is mainly mediated by numerous small molecular compounds [4]. Examples include dimethyl sulfoxide, benzoic acid, cinnamic acid, cinnamic aldehyde, eugenol, methyl niconate, and sorbic acid. Among fragrances, cinnamic aldehyde, cinnamic alcohol, isoeugenol, hydroxycitronellal, and geraniol can induce NICoU [5, 6, 7].

The mechanism of the NICoU is not well understood. There are at least three possible pathways or concepts to explain these reactions: nonspecific histamine release, modulation of the metabolism of arachidonic acid, and activation of cutaneous nerves including mediators such as substance P [8]. However antihistamines such as terfenadine do not inhibit most of the nonimmunological urticarial reactions, whereas nonsteroidal anti-inflammatory drugs (NSAIDS) can inhibit them after oral or topical application [9]. Interestingly UV light is capable to inhibit this reaction [4].

Although nonimmunological contact urticaria is more common than immunological contact

H.F. Merk
Department of Dermatology & Allergology,
RWTH Aachen University, Aachen, Germany

Dohlenfeld 8, 45479 Muelheim an der Ruhr,
Auf'm Hennekamp 50,
Düsseldorf D-40225, Germany
e-mail: hans.merk@post.rwth-aachen.de

© Springer International Publishing Switzerland 2018
J. Krutmann, H.F. Merk (eds.), *Environment and Skin*,
https://doi.org/10.1007/978-3-319-43102-4_6

Fig. 6.1 Basophil activation test (BAT) in a hairdresser with contact urticaria, eczema, and asthma after exposure to ammoniumpersulfate

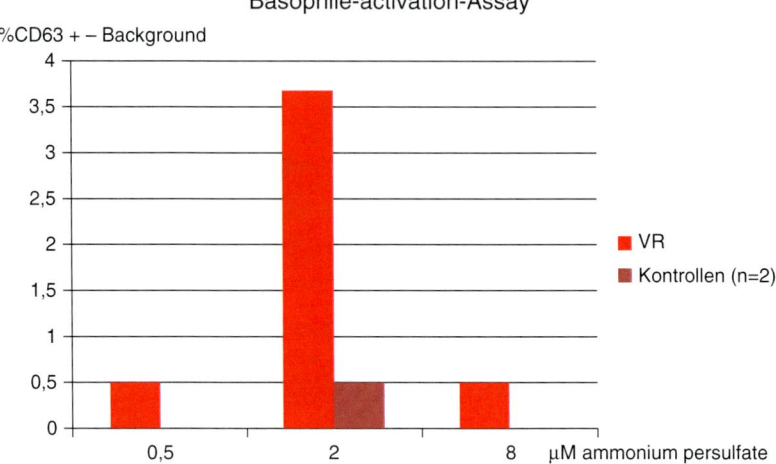

Basophile-activation-Assay

%CD63 + – Background

VR
Kontrollen (n=2)

0,5　　　　　　2　　　　　　8　　μM ammonium persulfate

Background: VR 1,46; Controls: 2,35/ 2,02; max stimulation: VR 68,29; Controls: 77,33/ 91,83

urticaria with regard to small molecular weight compounds, the latter form may have more serious consequences with regard to possible complications such as organ system involvement other than skin including anaphylaxis and death. Allergic contact urticaria is an immediate-type (type I) allergic reaction and in most cases IgE dependent. At least in cases of allergic reactions to small molecular weight compounds, both immediate-type and delayed-type reactions may exist. For example, in the case of a hairdresser with allergic reactions to ammonium persulfate, the patient first developed a delayed type of allergic reaction to ammonium persulfate, resulting in the precipitation of an allergic contact dermatitis, first as hand eczema, later spreading all over the entire body. Finally the patient developed also signs and symptoms of a contact urticaria and allergic asthma to ammonium persulfate and became positive to ammonium persulfate in the basophil activation test (BAT), which is in most cases an IgE-dependent activation (Fig. 6.1) [10].

Observation like this underlines that the skin is a preferred target organ for allergic reactions to small molecular compounds. Several skin diseases may be the result such as allergic contact dermatitis, allergic contact urticaria, and the multiple signs and symptoms of allergic drug reactions. More than this the skin seems to be the major organ which is able to initiate allergic reaction by sensitization of the immune system

Table 6.1 Small molecular compounds, which may elicit allergic contact urticaria, contact urticaria syndrome, and asthma

Epoxy resins
Persulfates
Chlorhexidine
Aziridine
Methacrylates
Ethanolamines
Formaldehyde
Colophonium
Isocyanates
Chloramine
ß-Lactam antibiotics
Cyclic acid anhydrides (e.g., methyl hexahydrophthalic anhydride), which is a constituent of epoxy resin
Bisphenol A (also in epoxy resins)
Metals (aluminum, chromium, cobalt, iridium salts, nickel, rhodium, platinum salts)

against such compounds [11]. In order to study possible allergic asthma to small molecular weight compounds in mice, it is necessary first to sensitize the animals by applying the compound of interest to the skin, and after developing an allergic response in the skin, it is possible to induce an asthma by inhalation—inhalation without further skin contact will not lead to an asthma response [12]. Most allergic contact urticarial reactions to small molecular compounds (Table 6.1) are occurring by occupa-

Table 6.2 Stages of the contact urticaria syndrome [13]

Stage 1	Localized urticaria (redness, swelling); immediate contact dermatitis (eczema-protein contact dermatitis)
	Itching, tingling, or burning sensation
Stage 2	Generalized urticaria
Stage 3	Multiple organ involvement [*rhinoconjunctivitis* (running nose, watery eyes), *bronchial asthma* (bronchospasmus, wheezing), *angioedema* and orolaryngeal symptoms (lip swelling, hoarseness, difficulty in swallowing), or *gastrointestinal symptoms* (nausea, vomiting, diarrhea, cramps)]
Stage 4	Anaphylactic shock

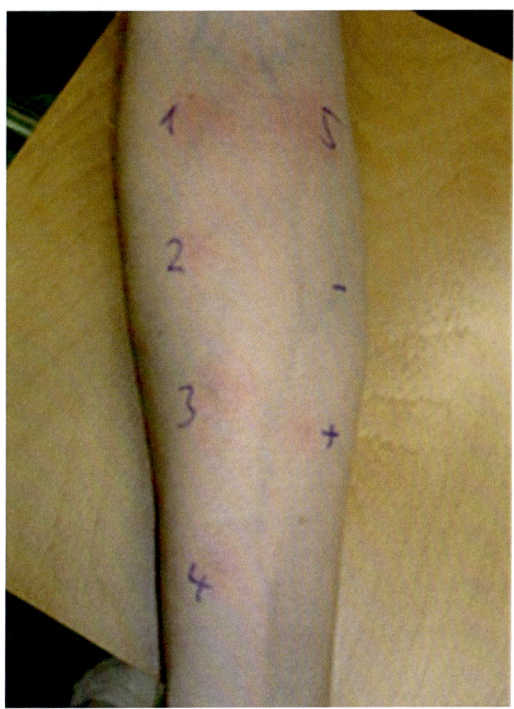

Fig. 6.2 Prick test (butcher with protein contact urticaria). *1* Beef, *2* Pig, *3* Chicken, *4* Turkey, *5* Lamb, *−* 0.9% NaCl, *+* histamine. Vanstreels L, Merk HF: Protein contact dermatitis in a butcher. Hautarzt 63 (2012) 926–8

tional exposure, and possible prick tests to confirm these allergic reactions have been recommended recently [14]. Benzophenone-3 with an absorption spectrum at 350 nm is incorporated in sunscreens and self-tanning products and can induce severe contact urticaria syndrome up to stage IV (Table 6.2) [5]. Similar reactions can be induced by menthol in toothpaste and in makeup remover [15].

There are many other examples of cosmetic compounds causing allergic and nonallergic contact urticaria including hair care products, antimicrobial agents and preservatives, sunscreens, fragrance compounds, toothpaste flavors, or plant-derived and animal-derived cosmetic ingredients [5]. Recently in particular wheat and hydrolysates were discussed, because they induce an exercised induced urticaria, angioedema, and even anaphylaxis after topical application [16, 17]. The reason is sensitizations to gliadins [18, 19, 20].

The majority of allergic contact urticaria or ICoU are mediated by proteins, in particular food components. In most cases they are IgE-dependent immunologically mediated reactions. It starts as a contact urticaria, and if the antigen exposure persists, it can develop a "protein contact dermatitis," most often localized on the hands and forearms and has been described as such first by Hjorth and Roed-Petersen [15]. It also can progress to a contact urticaria syndrome (Table 6.2) [3]. Also in cases of food allergy, the importance of epicutaneous sensitization of foods as inducer of those food allergy has been underlined recently since in most cases food allergy is preceded by contact

urticaria by occupational cooking and skin care treatment [21]. The protein contact dermatitis plays an important role in occupational dermatology and is precipitated in particular in butchers and slaughterhouse workers, those who sells vegetables and other food as well as those who are working in the kitchen of restaurants and hotels, and farmers; grains and enzymes can play a role in the protein contact dermatitis including contact urticaria syndrome in bakers and millers [22–24, 25] (Fig. 6.2). Nurses, doctors, as well as patients became sensitized to latex. It was possible to identify several components of latex which induced sensitization and allergic reactions subsequently, and to some extent the sensitization to some components was dependent from the kind of exposure [26]. In addition there are several cross-reactivities between latex and fruits such as banana and avocado [27].

Diagnostic tests to identify the culprit antigen include prick tests with commercial reagents and fresh material; also the "Reibtest" (gentle rubbing with the material) can be helpful in highly sensitized patients. Since these skin tests are possible also in cases of allergic reactions to proteins, they show that even proteins can be absorbed by the skin. Factors which reduce the barrier integrity and function of the skin include preexisting dermatitis such as irritant or atopic dermatitis, physical damage, chemical damage, e.g., by detergents, increased hydration, and occluded skin. In vitro tests can be applied to measure specific IgE in the serum, and the basophil activation test can be helpful, in particular if the antigen for the determination of antigens is not available for assays to detect specific IgE [28, 29, 30]. Specific IgE can be determined with the CAP-FEIA assay as well as with the technical platform of the microarray which in particular is helpful if one wants to determine protein or peptide components [27, 31]. In vitro tests are especially important and superior to skin tests, if the patient has a history of serious reactions such as anaphylaxis or angioedema or skin tests are not possible because of an acute eczema. In these cases in vitro tests are able to replace the skin tests.

References

1. Gimenez-Arnau A, Maurer M, De La Cuadra J, Maibach H. Immediate contact skin reactions, an update of contact urticaria, contact urticaria syndrome and protein contact dermatitis – "A Never Ending Story". Eur J Dermatol. 2010;20(5):552–62.
2. Wang CY, Maibach HI. Immunologic contact urticaria – the human touch. Cutan Ocul Toxicol. 2016;32:154–60.
3. Von Krogh C, Maibach HI. The contact urticaria syndrome. An update review. J Am Acad Dermatol. 1981;5:328–42.
4. Kim E, Maibach H. Contact urticarial. In: Greaves MW, Kaplan AP, editors. Urticaria and angioedema. New York, Basel: Marcel Dekker Inc.; 2004. p. 149–69.
5. Verhulst L, Goossens A. Cosmetic components causing contact urticaria: a review and update. Contact Dermatitis. 2016;75(6):333–44.
6. Gimenez-Arnau E. Chemical compounds as trigger factors of immediatw ontact skin reactions. In: Giménez-Arnau AM, Maibach HI, editors. Contact urticaria syndrome. 2nd ed. Boca Raton: CRC Press; 2015. p. 67–77.
7. Kireche M, Gimenez-Arnau E, Lepoittevin JP. Preservatives in cosmetics: reactivity of allergenic formaldehyde-releasers towards amino acids through breakdown products other than formaldehyde. Contact Dermatitis. 2010;63(4):192–20.
8. Kujala T, Lahti A. Duration of inhibition of non-immunologic immediate contact reactions by acetylsalicylic acid. Contact Dermatitis. 1989;21(1):60–1.
9. Lahti A, Väänänen A, Kokkonen EL, Hannuksela M. Acetylsalicylic acid inhibits non-immunologic contact urticaria. Contact Dermatitis. 1987;16(3):133–5.
10. Hoffmann HJ, Santos AF, Mayorga C, Nopp A, Eberlein B, Ferrer M, Rouzaire P, Ebo DG, Sabato V, Sanz ML, Pecaric-Petkovic T, Patil SU, Hausmann OV, Shreffler WG, Korosec P, Knol EF. The clinical utility of basophil activation testing in diagnosis and monitoring of allergic disease. Allergy. 2015;70(11):1393–40.
11. North CM, Ezendam J, Hotchkiss JA, Maier C, Aoyama K, Enoch S, Goetz A, Graham C, Kimber I, Karjalainen A, Pauluhn J, Roggen EL, Selgrade M, Tarlo SM, Chen CL. Developing a framework for assessing chemical respiratory sensitization: a workshop report. Regul Toxicol Pharmacol. 2016;80:295–309.
12. Pauluhn J. Development of a respiratory sensitization/elicitation protocol of toluene diisocyanate (TDI) in Brown Norway rats to derive an elicitation-based occupational exposure level. Toxicology. 2014;319:10–22.
13. Gimenez-Arnau A. Contact urticaria and the environment. Rev Environ Health. 2014;29:207–15.
14. Helaskoski E, Suojalehto H, Kuuliala O, Aalto-Korte K. Prick testing with chemicals in the diagnosis of occupational contact urticarial and respiratory diseases. Contact Dermatitis. 2014;72:20–32.
15. Giménez-Arnau AM, Maibach HI. Contact urticaria syndrome. 2nd ed. Boca Raton: CRC Press; 2015.
16. Fukutomi Y, Itagaki Y, Taniguchi M, Saito A, Yasueda H, Nakazawa T, Hasegawa M, Nakamura H, Akiyama K. Rhinoconjunctival sensitization to hydrolyzed wheat protein in facial soap can induce wheat-dependent exercise-induced anaphylaxis. J Allergy Clin Immunol. 2011;127(2):531–53.
17. Fukutomi Y, Taniguchi M, Nakamura H, Akiyama K. Epidemiological link between wheat allergy and exposure to hydrolyzed wheat protein in facial soap. Allergy. 2014;69(10):1405–11.
18. Adachi R, Nakamura R, Sakai S, Fukutomi Y, Teshima R. Sensitization to acid-hydrolyzed wheat protein by transdermal administration to BALB/c mice, and comparison with gluten. Allergy. 2012;67(11):1392–9.
19. Leheron C, Bourrier T, Albertini M, Giovannini-Chami L. Immediate contact urticaria caused by hydrolysed wheat proteins in a child via maternal skin contact sensitization. Contact Dermatitis. 2013;68(6):379–80. doi:10.1111/cod.12046.
20. Nakamura R, Nakamura R, Sakai S, Adachi R, Hachisuka A, Urisu A, Fukutomi Y, Teshima R. Tissue transglutaminase generates deamidated epitopes on gluten, increasing reactivity with hydrolyzed wheat protein-sensitized IgE. J Allergy Clin Immunol. 2013;132(6):1436.

21. Inomata N, Nagashima M, Hakuta A, Aihara M. Food allergy preceded by contact urticaria due to the same food: involvement of epicutaneous sensitization in food allergy. Allergol Int. 2015;64(1):73–8.

22. Amaro C, Goossens A. Immunological occupational contact urticaria and contact dermatitis from proteins: a review. Contact Dermatitis. 2008;58(2):67–75.

23. Doutre MS. Occupational contact urticaria and protein contact dermatitis. Eur J Dermatol. 2005; 15(6):419–24.

24. Vanstreels L, Merk HF. Protein contact dermatitis in a butcher. Hautarzt. 2012;63(12):926–8.

25. Lukacs J, Schliemann S, Elsner P. Occupational contact urticaria caused by food – a systematic clinical review. Contact Dermatitis. 2016;75(4):195–204.

26. Ott H, Schröder C, Raulf-Heimsoth M, Mahler V, Ocklenburg C, Merk HF, Baron JM. Microarrays of recombinant Hevea brasiliensis proteins: a novel tool for the component-resolved diagnosis of natural rubber latex allergy. J Investig Allergol Clin Immunol. 2010;20(2):129–38.

27. Ott H, Baron JM, Heise R, Ocklenburg C, Stanzel S, Merk HF, Niggemann B, Beyer K. Clinical usefulness of microarray-based IgE detection in children with suspected food allergy. Allergy. 2008;63(11):1521–8.

28. Erdmann SM, Heussen N, Moll-Slodowy S, Merk HF, Sachs B. CD63 expression on basophils as a tool for the diagnosis of pollen-associated food allergy: sensitivity and specificity. Clin Exp Allergy. 2003;33(5):607–14.

29. Erdmann SM, Sachs B, Schmidt A, Merk HF, Scheiner O, Moll-Slodowy S, Sauer I, Kwiecien R, Maderegger B, Hoffmann-Sommergruber K. In vitro analysis of birch-pollen-associated food allergy by use of recombinant allergens in the basophil activation test. Int Arch Allergy Immunol. 2005;136(3):230–8.

30. Vanstreels L, Merk HF. Value of in-vitro diagnostic tools after anaphylaxis. Hautarzt. 2013;64(2):93–6.

31. Ott H, Lehmann S, Wurpts G, Merk HF, Viardot-Helmer A, Rietschel E, Baron JM. Anaphylaxis in childhood and adolescence. Hautarzt. 2007;58(12):1032–40.

Risk Assessment for Contact Allergens

7

David A. Basketter

7.1 Introduction

Our world is formed entirely from chemicals, the great majority of them delivering vital properties—nutrition, homes, transport, medicines, hygiene and so on. However, some of these chemicals, as well as others in the environment, also have other capacities, including the ability to combine with our proteins and thereby cause allergy. The underlying mechanisms associated with this are reviewed elsewhere in this book and will not be repeated here (see Chap. 10). This section will focus wholly on the assessment of the risk presented to human health. Before that is started though, some definitions are necessary:

A *skin sensitiser* is a protein reactive chemical which can induce a state of delayed-type, cell-mediated hypersensitivity, which in humans is termed *contact allergy*.

Contact allergy can be detected by a diagnostic patch test with the offending chemical. If there is a positive reaction, the individual is *sensitised*.

The skin disease *allergic contact dermatitis* will be elicited in a *sensitised* individual where there is sufficient dermal exposure.

Hazard identification and characterisation involves determining whether a particular chemical possesses skin-sensitising properties and, if it does, measuring the strength of that property, commonly referred to as skin-sensitising potency.

Risk assessment requires the combination of sensitising potency with dermal exposure to that skin sensitiser so that the risk to human health can be managed.

The processes associated with the risk assessment of contact allergens have been reviewed in several recent occasions [1, 2], so a primary focus in the present material will be on the more recent developments, as well as on how the feedback from dermatology clinics informs us on the performance (i.e. success or failure) of the risk assessment.

7.2 Hazard Identification and Characterisation

Predictive test methods and their protocols, advantages and disadvantages generally fall outside the scope of the material here, with thorough reviews being readily available which document progress over several decades (e.g. [3, 4]). Nevertheless, some key points must be addressed.

The earliest predictive assays for the identification of chemical skin sensitisation hazards involved the use of a range of guinea pig models (detailed in [3]). All employed a variety of procedures over a period of 2–3 weeks which endeavoured to induce sensitisation to the test chemical. The success of

D.A. Basketter
DABMEB Consultancy Ltd.,
Sharnbrook MK44 1PR, UK
e-mail: dabmebconsultancyltd@me.com

© Springer International Publishing Switzerland 2018
J. Krutmann, H.F. Merk (eds.), *Environment and Skin*,
https://doi.org/10.1007/978-3-319-43102-4_7

57

this was then revealed 1–2 weeks later by dermal challenge with the same substance. Test concentrations depended on the irritancy potential of the substance, with the consequence that irritant substances would be evaluated at much lower concentrations than those which were less irritating to the skin. In addition, the responses elicited on the skin were assessed by a nonstandardised and subjective visual scoring. Thus, although the methods were often very effective in the identification of skin sensitisation hazards, they were much less well suited to the measurement of the potency of an identified sensitiser [5–7].

In contrast to the guinea pig methods, the murine local lymph node assay (LLNA) determines the effectiveness of the topical induction regime via direct measurement of the induced proliferation in draining lymph nodes [8, 9]. Thus, with an objective and quantitative endpoint, not only can potential skin-sensitising chemicals be identified, there is also the opportunity to make further use of this data for potency assessment [10]. The details of how this is done are now long known [11]. More importantly, large datasets of skin-sensitising chemicals and their relative potency have been published [12, 13]. However, the most important point is whether and to what extent murine predictions of relative sensitising potency reflect the situation in the species which concerns us. As might be anticipated, the mouse is a useful, but not a perfect, predictor for humans, plus the EC3 value is a biological measure, subject to the usual vagaries [14]. This review notes that EC3 values cover at least a semilog range of uncertainty. It is also pertinent to ask how well it correlates to known HRIPT NESIL values derived by experiment. Human information on the intrinsic relative skin-sensitising potency of chemicals is limited, but nevertheless, several publications over the last decade or so have addressed this question ([6, 15–17]). Using this type of data, the subject has been reviewed [6, 15–20]. The conclusion is that the predictions are very helpful but that some substances seem substantially discordant (salicylates form one example). It is pertinent to be reminded that predictive toxicology assays are never perfect [21].

7.3   Risk Assessment

Historically, risk assessment for contact allergens did not follow classic toxicological principles, i.e. establishment of a no-effect level (NOEL), application of safety factors to derive an acceptable human exposure limit and margin of exposure calculation. Sometimes the terminology varies, but the aim is to ensure that the daily human exposure level does not exceed an experimentally determined safe limit. In contrast, skin sensitisers, having been identified in guinea pig testing, were assessed only by a comparative, often very conservative, process. Details of the strategy were not often published, but examples do exist (e.g. [22, 23]). In some cases, the conclusion of a risk assessment was "confirmed" by the conduct of a human repeated insult patch test (HRIPT)—a very dubious process, both from an ethical and a scientific perspective [24]. The human skin exposure element of the risk assessment was largely eliminated by the strategy adopted, since at the core of this approach was the comparison of the sensitising activity, in a guinea pig assay, of a new ingredient and an old ingredient. The latter must be used in the same product type and at a similar concentration and benefit from acceptability in the marketplace which has been convincingly established over a period of many years. So, in the classic toxicology equation "risk = hazard x exposure", this comparative approach kept the exposure part constant and the hazards compared so that the risk remained acceptable. However, the intrinsic variability of guinea pig assays meant that this comparison could only be done reliably within specific testing institutions. Furthermore, where an ingredient was designated for a new product type or where there was no ingredient to permit historic comparisons to be made, then the risk assessor faced an impasse. In such situations, the only course of action generally is precautionary, which obviously limits the risk of ACD but could also fail to enable the use of valuable new substances.

In the search for a new, transparent and more flexible approach to the risk assessment of contact allergens, two new aspects of thinking needed to coincide, quantitation of dermal exposure and measurement of relative sensitising potency. The determination of the relative potency of a skin-sensitising chemical has been discussed in the preceding section. This is done using the LLNA and is known as the EC3 value (the estimated concentration to cause threefold stimulation). EC3 values for several hundred sensitising chemicals have been published [12, 13]. EC3 values also are robust within and between laboratories [14]. Clearly, the comparative approach noted above for guinea pig assays also can be applied successfully here, since replacement of an ingredient in a formulation with one having a similar (or better) EC3 value (and assuming a similar use concentration) is reasonable.

However, completion of a novel and transparent risk assessment necessitates integration of potency information with knowledge of human skin exposure. In reality, the most potent contact allergen can be used safely if exposure is sufficiently low, and conversely, even a very weak allergen can be a problematic cause of ACD if skin exposure is too high. The key question therefore is this: "How can the information on potency and exposure be integrated?". Following the general principles of toxicological risk assessment, the LLNA EC3 value is translated into a skin sensitisation induction threshold,

expressed in dose per unit area (normally µg/cm^2), to which safety factors are applied such that an acceptable exposure level (AEL) can be determined, similar to the acceptable daily intake (ADL) figure associated with repeated dose toxicity. An overview of the approach is presented in Fig. 7.1. Note that the experimental threshold derived from the EC3 value is termed the "no expected sensitisation induction level (NESIL)", and remember that this corresponds to the induction threshold in a 100-subject human repeated insult patch test (HRIPT), in reality a human sensitisation study, not normally done in the twenty-first century except to confirm the absence of sensitising effect. Complete practical details and scientific support for this process, termed quantitative risk assessment (QRA), have been published [25–27]. In addition, several practical examples of the use of QRA have been presented, involving a range of well-known human contact allergens including fragrances [19, 20, 25, 27–29]; the transition metals nickel, chromate and cobalt; and several preservatives [30, 31]. The QRA has also been modified to take into account exposure at mucosal surfaces [32]. Most recently, it has been updated to address aggregate exposure and the latest scientific knowledge [33].

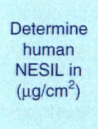

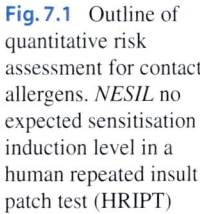

Fig. 7.1 Outline of quantitative risk assessment for contact allergens. *NESIL* no expected sensitisation induction level in a human repeated insult patch test (HRIPT)

| Determine human NESIL in (µg/cm^2) | Add 1–10× safety factor for human variability | Add 0.3–1× safety factor for vehicle matrix | Add 1–3× safety factor for frequency & duration variables | Add 1–10× safety factor for skin condition | Acceptable exposure level (µg/cm^2) is compared to aggregate exposure level (µg/cm^2) |

Table 7.1 An example of how human potency categories may indicate the NESIL

LLNA EC3 value (%)	Human category[a]	Default NESIL (µg/cm[b])[b]
<0.02	1	10
<0.20	2	100
<2.0	3	1000
<20	4	10,000
<100	5	100,000
Negative	6	1,000,000

[a]As defined in [19, 20]
[b]Modifed from LLNA EC3% values can be converted to µg/cm[b] figures via multiplication by a factor of 250

Central to QRA for contact allergens are three elements, the determination of the NESIL, the application of appropriate safety factors and the estimation of skin exposure. Uncertainties exist with all of these. For example, the NESIL is derived most commonly from the LLNA EC3 value [25, 34, 35]. However, it is known that although useful, this may not always be a precise predictor of relative human potency. One strategy in which use of the LLNA EC3, together with any other information, is used to allocate a substance to one of the six human potency categories has been proposed [19, 20]. These would then lead to default NESIL values, rather than a direct one-to-one translation from the murine EC3 value, which may be more appropriately conservative. The detail of the approach is shown in Table 7.1 and represents an evolution of a strategy that was first suggested more than a decade ago [27].

The safety factors noted in Fig. 7.1, often more accurately referred to as uncertainty factors, represent expert judgement more than anything else. Probably the most soundly based on real data is the exposure matrix factor where information on the impact of vehicle has been published and indicates that allowance for a ten-fold variation is reasonable [36]. The human variability factor of 10 is in addition to the heterogeneity intrinsic in the 100 volunteers participating in the HRIPT. There are surely those who will feel that this is insufficient, but there is no doubt that it is consistent with historical practice in toxicological risk assessment [37]. The final factor is directed towards taking account of expo-sure considerations that are above and beyond those that already are built into the calculation of the dermal exposure dose. Thus, considerations such as chronic exposure and/or the presence of skin inflammation can be accommodated. In practice, values of 1, 3 or 10 are selected, using all available information together with the knowledge of how and where skin exposure to the sensitising chemical will occur.

7.4 Contact Allergy

Uniquely in toxicology, this paradigm of skin sensitisers, contact allergy and allergic contact dermatitis represents a feedback loop of real substance (Fig. 7.2). In fact, allergic contact dermatitis to sensitising chemicals represents perhaps the most common expression of immunotoxicological disease: the asymptomatic sensitised state, contact allergy, can be found in at least 20% of adults [38], and all of these are at risk of the expression of allergic eczema if they have sufficient skin contact with the substance to which they are contact allergic. The actual frequency, the epidemiology, of allergic contact dermatitis in the general population is not well defined, but we know that there is a problem, since dermatology clinics around the world diagnose this disease in many patients every week. This diagnosis depends largely on two things, the anamnesis and the diagnostic patch test [39]. The latter, the patch test, applies suspect sensitising substances to the back of the patient for 48 hours and then examines the skin over the next days for the expression of allergic contact dermatitis at the site of contact. The importance of the data generated in this way should not be underestimated [40, 41]. It provides the feedback loop which informs us whether and to what extent the risk assessment (and risk management) has been successful. So far, the only detailed retrospective analysis has been completed for three fragrance ingredients, which suggests that had QRA been applied, the burden of allergic contact dermatitis to these substances would have been reduced compared to the level actually experienced [40].

Fig. 7.2 Risk
assessment and clinical
feedback loop

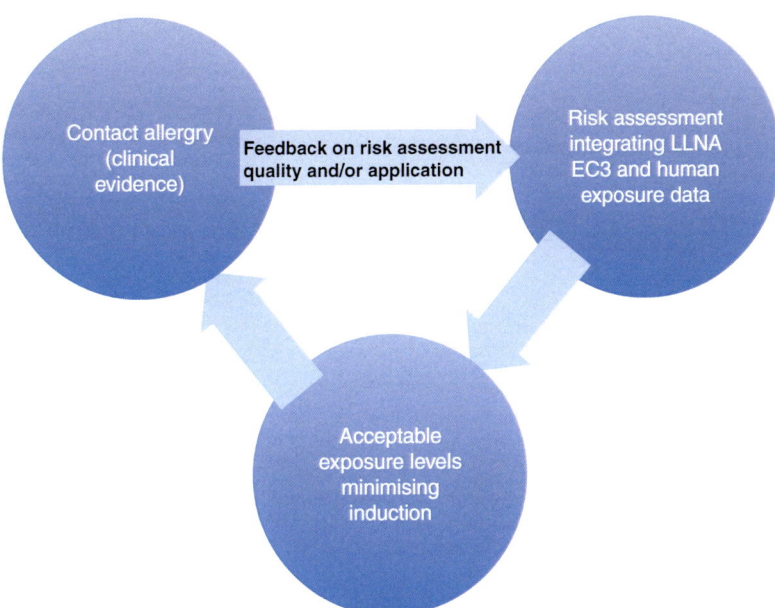

Fig. 7.2 Risk assessment and clinical feedback loop

Now, there are several thousand substances that have been identified clinically as potential contact allergens. There are many others noted in regulatory databases (e.g. [42]). The reality is that most people, most of the time, do come into contact with skin-sensitising substances and can do so with equanimity—contact allergy/allergic contact dermatitis does not result. Cosmetic products have, for many years, been one of the most important causes of this disease [43], but it remains the case that the great majority of people use multiple cosmetic products daily and do so in complete safety. Whether the balance between safety and commercial interests is right is beyond the scope of this chapter, but it is instructive to examine a recent case where matters have not gone according to plan and resulted in an epidemic of allergic contact dermatitis.

Methylisothiazolinone (MI) is a preservative which was found useful in many industrial applications and in more recent years in cosmetics. It had long been known to be a skin sensitiser [44]. Its activity in this respect had also been assessed in the LLNA, where it was identified as a strong sensitiser, similar in potency to formaldehyde ([45]). Thus it could reasonably be argued that the risk was known and tools for its assessment were avail-

able, such that human health could be protected. The specific risk in relation to cosmetic use was recognised and an upper limit of 100 ppm in products applied in Europe [46, 47]. What followed was a relatively sudden epidemic, whose scale eventually reached proportions generally unseen before [48, 49]. Aspects of this failure to protect human health have been reviewed recently and will not be reworked in detail here [50]. Nevertheless, several key learnings can be derived:

- Contact allergy risk assessment only works if it is used.
- In vivo measures of sensitising potency are useful but imperfect.
- Risk assessment needs to take account of aggregate exposure.
- Clinical feedback requires prompt, coordinated action.

These may not be the only things that can be learned, but they are central to making sure that the risk assessment for contact allergens is undertaken in an optimal fashion. It seems self-evident that risk assessment only works if it is used, but a central problem with MI has been allergic contact dermatitis to cosmetic products in which it has been used as a preservative and where it is fair to

Table 7.2 QRA applied to methylisothiazolinone

Product type	Original EU generic limit[a]	QRA-derived product limit[b]	QRA-based EU generic limit[c]
Non-aerosol deodorant		5	
Face cream/body lotion	100 ppm	10	5 ppm
Liquid soap		15	
Shampoo	100 ppm	150	15 ppm

[a]As defined by the EU in 2003
[b]This is the maximum acceptable level of MI that would have been allowed if the published QRA calculation had been followed, using a NESIL of 15 µg/cm[b] [30]
[c]Based on the QRA calculations and given that the EU sets limit only for generic leave-on and rinse-off products, the lowest figure for each product category would be used

suspect that proper risk assessment was not done. Many, perhaps most, companies relied on the EU approval for use at up to 100 ppm in such products [46, 47]. Here, all available data was used to complete the toxicological evaluation and thereby recommend a safe use level. However, QRA was new and was not used, but neither was MI for some time—the epidemic only became apparent in perhaps 2011/2012. By that time, QRA use had been described in detail and worked examples provided, including for preservatives in cosmetics [30, 31]. Table 7.2 shows what the impact use of QRA would have had—the limits are generally lower than the generic EU value. Thus, it can be argued that the EU regulatory limit for cosmetics might have been high but perhaps a key failure occurred in companies, which adopted the regulatory limit, rather than conducting their own safety evaluation using the latest contact allergy QRA.

A second possible source of failure, which is generic to the currently published QRA, is that the exposure calculations are completed on a single product basis [25]. However, it is reasonable to anticipate that at least some individuals will experience exposure to MI from multiple sources, even within the cosmetic/personal care category. A programme of work is underway to take account of aggregate exposure in contact allergy risk assessment (e.g. [51–55]). Although these references all refer to cosmetics, the principles therein are broadly applicable to skin contact sources.

The third failure for MI was in not taking account with sufficient alacrity of clinical data indicating that MI was causing contact allergy/allergic contact dermatitis [48]. This produces what is known as the Dillarstone phenomenon, where ignoring of early signals of preservative contact allergy means that an epidemic occurs and which is only latterly mitigated by delayed action [56]. The importance of the clinical contact allergy information has recently been re-emphasised [41]. Having said this, it is nevertheless only fair to note that action on MI use in cosmetics was proposed by European industry [57], well ahead of action via regulation in that region, which at the time of writing is still awaited, despite expert opinion [58].

7.5 Allergic Contact Dermatitis

Allergic contact dermatitis is the eczema which develops when an individual with contact allergy to a sensitising chemical has sufficient skin exposure to that substance. Thus, the prevalence/absence of this disease in general and/or occupational populations provides a powerful indicator of the success of contact allergy risk assessment and any consequent risk management measures. That allergic contact dermatitis represents one of the most important causes of occupational skin disease is a testament to the continuing failure to properly assess and manage the risks [59–61]. Similarly, the high levels of allergy to fragrances and preservatives, especially in cosmetics, require positive action [48, 62].

As indicated in the section above, when contact allergy risk assessment goes wrong, allergic contact dermatitis is the result. This brings an entirely new dimension to safety evaluation, since the skin exposure levels which are sufficient to avoid the induction of contact allergy may be inadequate to avoid the elicitation of eczema in a sensitised individual. Elicitation thresholds typically are much lower, and whilst it is of course possible to undertake clinical studies in allergic subjects to identify safe use levels, the prediction of these from toxicological data, such as the LLNA EC3 value, is generally felt not to be possible [5]. As a consequence, the strategy implemented by regulatory bodies often is to adopt pragmatic values, typically 100 ppm for rinse-off products and 10 ppm for leave-on products (e.g. [63]). There has been a meta-analysis of clinical elicitation in dose-response data which provides some support for this type of approach [64]. However, ideally, clinical studies should be conducted, such that more appropriate allergen-specific thresholds can be found, typically by using a repeated open application test [65–67]. Note here that the aim might either be to limit dermal exposure to provide direct protection, such as with the European nickel regulation [68]. Alternatively, it might be simply to provide a basis on which to define a concentration limit that triggers an appropriate product warning label, such as with the EU cosmetic and detergent regulations regarding fragrance allergens [46, 47, 63].

Conclusions

Skin sensitising chemicals are common in our general and occupational environment, such that all of us have daily exposure to multiple substances with this property, often repeatedly within a 24h period. That the majority do not experience allergic contact dermatitis testifies to the risk assessment (and consequent risk management) that is undertaken. However, it also testifies to the reality that exposure is relatively low for many sensitisers and humans are probably not that easy to sensitise (i.e. make contact allergic). That said, there is also a burden of allergic contact dermatitis, both occupationally and in general consumers, that remains far higher than it should. Widespread and effective implementation of thorough contact allergy risk assessment is the key to reducing the morbidity of this disease.

References

1. Honda T, Egawa G, Grabbe S, et al. Update of immune events in the murine contact hypersensitivity model: toward the understanding of allergic contact dermatitis. J Invest Dermatol. 2013;133:303–15.

2. Martin SF. New concepts in cutaneous allergy. Contact Dermatitis. 2015;72:2–10.
3. Andersen KE, Maibach HI. Current problems in dermatology 14: contact allergy predictive tests in guinea pigs. Basel: Karger; 1985.
4. Thyssen JP, Giménez-Arnau E, Lepoittevin JP, et al. The critical review of methodologies and approaches to assess the inherent skin sensitization potential (skin allergies) of chemicals. Part I. Contact Dermatitis. 2012;66(Suppl 1):11–24.
5. Basketter DA, Andersen KE, Lidén C, et al. Evaluation of the skin sensitising potency of chemicals using existing methods and considerations of relevance for elicitation. Contact Dermatitis. 2005;52:39–43.
6. Basketter DA, Clapp C, Jefferies D, et al. Predictive identification of human skin sensitisation thresholds. Contact Dermatitis. 2005;53:260–7.
7. van Loveren H, Cockshott A, Gebel T, et al. Skin sensitization in chemical risk assessment: report of a WHO/IPCS international workshop focusing on dose-response assessment. Regul Toxicol Pharmacol. 2008;50:155–99.
8. Gerberick GF, Ryan CA, Kimber I, et al. Local lymph node assay: validation assessment for regulatory purposes. Am J Contact Dermat. 2000;11:3–18.
9. Kimber I, Basketter DA. The murine local lymph node assay; collaborative studies and new directions: a commentary. Food Chem Toxicol. 1992;30:165–9.
10. Kimber I, Basketter DA. Contact sensitization: a new approach to risk assessment. Hum Ecol Risk Assess. 1997;3:385–95.
11. Basketter DA, Lea L, Cooper K, et al. A comparison of statistical approaches to derivation of EC3 values from local lymph node assay dose responses. J Appl Toxicol. 1999;19:261–6.
12. Gerberick GF, Ryan CA, Kern PS, et al. Compilation of historical local lymph node data for evaluation of skin sensitization alternative methods. Dermatitis. 2005;16:157–202.
13. Kern PS, Gerberick GF, Ryan CA, et al. Historical local lymph node data for the evaluation of skin sensitization alternatives: a second compilation. Dermatitis. 2010;21:8–32.
14. Basketter DA, Gerberick GF, Kimber I. The local lymph node assay EC3 value: status of validation. Contact Dermatitis. 2007;57:70–5.
15. Basketter DA, McFadden JP. Cutaneous allergies. In: Dietert RR, Luebke RW, editors. Immunotoxicity, immune dysfunction and chronic disease. New York: Humana Press; 2012. p. 103–26.
16. Ryan CA, Gerberick GF, Cruse LW, et al. Activity of human contact allergens in the murine local lymph node assay. Contact Dermatitis. 2000;43:95–102.
17. Schneider K, Akkan Z. Quantitative relationship between the local lymph node assay and human skin sensitization assays. Regul Toxicol Pharmacol. 2004;39:245–55.
18. Api AM, Basketter DA, Lalko J. Correlation between experimental human and murine skin sensitization induction thresholds. Cut Ocul Toxicology. 2014;28:1–5.

19. Basketter DA, Alepee N, Ashikaga T, et al. Categorisation of chemicals according to their relative human skin sensitizing potency. Dermatitis. 2014;25:11–21.

20. Basketter DA, Lemoine S, McFadden JP. Skin sensitisation to fragrance ingredients: is there a role for household cleaning/maintenance products? Eur J Dermatol. 2015;25(1):7–13.

21. Basketter DA, McFadden JF, Gerberick F, et al. Nothing is perfect, not even the local lymph node assay: a commentary and the implications for REACH. Contact Dermatitis. 2009;60:65–9.

22. Basketter DA, Gerberick GF, Robinson M. Risk assessment. In: Kimber I, Maurer T, editors. The toxicology of contact hypersensitivity. London: Taylor and Francis; 1996. p. 152–64.

23. Robinson MK, Nusair TL, Fletcher ER, et al. A review of the Buehler guinea pig skin sensitization test and its use in a risk assessment process for human skin sensitization. Toxicology. 1990;61:91–107.

24. Basketter DA. The human repeated insult patch test in the 21st century: a commentary on ethics and validity. Cut Ocul Toxicol. 2009;28:49–53.

25. Api AM, Basketter DA, Cadby PA, et al. Dermal sensitization quantitative risk assessment (QRA) for fragrance ingredients. Regul Toxicol Pharmacol. 2008;52:3–23.

26. Felter SP, Ryan CA, Basketter DA, et al. Application of the risk assessment paradigm to the induction of allergic contact dermatitis. Regul Toxicol Pharmacol. 2003;37:1–10.

27. Gerberick GF, Robinson MK, Felter S, et al. Understanding fragrance allergy using an exposure-based risk assessment approach. Contact Dermatitis. 2001;45:333–40.

28. Api AM, Vey M. Implementation of the dermal sensitization quantitative risk assessment (QRA) for fragrance ingredients. Regul Toxicol Pharmacol. 2008;52:53–61.

29. Corea N, Basketter DA, van Asten A, et al. Fragrance allergy: assessing the risk from fabric washing products. Contact Dermatitis. 2006;55:48–53.

30. Basketter DA. Methyldibromo glutaronitrile, skin sensitisation and quantitative risk assessment. Cut Ocul Toxicol. 2010;29:4–9.

31. Basketter DA, Clapp CJ, Safford BJ, et al. Preservatives and skin sensitisation quantitative risk assessment: risk benefit considerations. Dermatitis. 2008;19:20–7.

32. Farage MA, Bjerke DL, Mahony C, et al. Quantitative risk assessment for the induction of allergic contact dermatitis: uncertainty factors for mucosal exposures. Contact Dermatitis. 2003;49:140–7.

33. Basketter DA, Safford RJ. Skin sensitisation quantitative risk assessment; a review of underlying assumptions. Regul Toxicol Pharmacol. 2016;74:105–16.

34. Safford RJ. The dermal sensitisation threshold- a TTC approach for allergic contact dermatitis. Regul Toxicol Pharmacol. 2008;51:195–200.

35. Safford RJ, Aptula AO, Gilmour N. Refinement of the dermal sensitisation threshold (DST) approach using a larger dataset and incorporating mechanistic chemistry domains. Regul Toxicol Pharmacol. 2011;60:218–24.

36. Jowsey IR, Clapp CJ, Safford B, et al. The impact of vehicle on the relative potency of skin sensitising chemicals in the local lymph node assay. Food Chem Toxicol. 2008;27:67–75.

37. Felter SP, Robinson MK, Basketter DA, et al. A review of the scientific basis for uncertainty factors for use in quantitative risk assessment for the induction of allergic contact dermatitis. Contact Dermatitis. 2002;47:257–66.

38. Thyssen JP, Linneberg A, Menné T, et al. The epidemiology of contact allergy in the general population-prevalence and main findings. Contact Dermatitis. 2007;57:287–99.

39. Lindberg M, Matura M. Chapter 13: Patch testing. In: Johansen JD, Frosch PF, Lepoittevin JP, editors. Contact dermatitis. 5th ed. Berlin: Springer; 2011. p. 439–64.

40. Marie Api A, Belsito D, Bickers D. Quantitative risk assessment of contact sensitization: clinical data to assess utility of the model. Dermatitis. 2010;21:207–13.

41. Basketter DA, White IR. Diagnostic patch testing – does it have a wider relevance? Contact Dermatitis. 2012;67:1–2.

42. ECHA. European chemicals agency classification and labelling inventory. 2015. http://echa.europa.eu/web/guest/information-on-chemicals/cl-inventory-database. Accessed 14 Jan 2015.

43. Alani JI, Davis MD, Yiannias JA. Allergy to cosmetics: a literature review. Dermatitis. 2013;24:283–90.

44. Bruze M, Fregert S, Gruvberger B, et al. Contact allergy to the active ingredients of Kathon CG in the guinea pig. Acta Derm Venereol. 1987;67:315–20.

45. Basketter DA, Kimber I. Chapter 13: Predictive tests for irritants and allergens and their use in quantitative risk assessment. In: Johansen JD, Frosch PF, Lepoittevin JP, editors. Contact dermatitis. 5th ed. Berlin: Springer; 2011. p. 229–40.

46. EU. The scientific committee on cosmetic products and non-food products intended for consumers. Opinion concerning methylisothiazolinone. 2003. http://ec.europa.eu/food/fs/sc/sccp/out_201.pdf. Accessed 14 Jan 2015.

47. EU. The European Parliament and the Council of the European Union. Directive 2003/15/EC of the European Parliament and of the Council of 27 February 2003 amending Council Directive 76/768/EEC on the approximation of the laws of the Member States relating to cosmetic products. 2003. http://eurlex.europa.eu/LexUriServ/LexUriServ.do?uri=OJ:L:2003:066:0026:0035:en:PDF. Accessed 16 Dec 2013.

48. Gonçalo M, Goossens A. Whilst Rome burns: the epidemic of contact allergy to methylisothiazolinone. Contact Dermatitis. 2013;68:257–8.

49. Lundov MD, Opstrup MS, Johansen JD. Methylisothiazolinone contact allergy: a growing epidemic. Contact Dermatitis. 2013;69:271–5.

50. Basketter DA, White IR, McFadden JP, et al. Skin sensitization: integration of clinical data into hazard identification and risk assessment. Human Exp Toxicol. 2015;34(12):1222–30.

51. Crème RIFM. 2014. http://www.Cremeglobal.Com/Modelling-software/creme-care-cosmetics/creme-rifm. Accessed 18 Jan 2015.

52. Crème Global. Aggregate exposure from real consumer data. 2014. http://www.cremeglobal.com/modelling-software/creme-care-cosmetics/. Accessed 18 Jan 2015.

53. Hall B, Tozer S, Safford B, et al. European consumer exposure to cosmetic products, a framework for conducting population exposure assessments. Food Chem Toxicol. 2007;45:2097–108.

54. Hall B, Steiling W, Safford B, et al. European consumer exposure to cosmetic products, a framework for conducting population exposure assessments part 2. Food Chem Toxicol. 2011;49:408–22.

55. McNamara C, Rohan D, Golden D, et al. Probabilistic modelling of European consumer exposure to cosmetic products. Food Chem Toxicol. 2007;45:2086–96.

56. Dillarstone A. Cosmetic preservatives. Contact Dermatitis. 1997;37:190.

57. Cosmetics Europe. Cosmetics Europe Recommendation on MIT. 2013. https://www.cosmeticseurope.eu/news-a-events/news/647-cosmetics-europe-recommendation-on-mit.html. Accessed 18 Jan 2015.

58. EU. Scientific committee on consumer safety opinion on methylisothizzolinone. Adopted 12 December 2013 and revised on 27 March 2014. 2014. http://ec.europa.eu/health/scientific_committees/consumer_safety/docs/sccs_o_145.pdf. Accessed 18 Jan 2015.

59. Anderson SE, Meade BJ. Potential health effects associated with dermal exposure to occupational chemicals. Environ Health Insights. 2014;8(Suppl 1):51–62.

60. Basketter DA. Skin sensitization: strategies for the assessment and management of risk. Br J Dermatol. 2008;159:267–73.

61. Holness DL. Occupational skin allergies: testing and treatment (the case of occupational allergic contact dermatitis). Curr Allergy Asthma Rep. 2014;14:410.

62. Cheng J, Zug KA. Fragrance allergic contact dermatitis. Dermatitis. 2014;25:232–45.

63. EU. European Detergents Regulation (EC) No. 648/2004 as amended 14/03/2012. 2012. http://eur-lex.europa.eu/LexUriServ/LexUriServ.do?uri=CONSLEG:2004R0648:20120419:EN:PDF. Accessed 7 Jun 2013.

64. Fischer LA, Menné T, Voelund A, et al. Can exposure limitations for well-known contact allergens be simplified? An analysis of dose-response patch test data. Contact Dermatitis. 2011;64:337–42.

65. Fischer LA, Johansen JD, Menné T. Nickel allergy: relationship between patch test and repeated open application test thresholds. Br J Dermatol. 2007;157:723–9.

66. Fischer LA, Johansen JD, Menné T. Methyldibromoglutaronitrile allergy: relationship between patch test and repeated open application test thresholds. Br J Dermatol. 2008;159:1138–43.

67. Schnuch A, Uter W, Dickel H, et al. Quantitative patch and repeated open application testing in hydroxy-isohexyl 3-cyclohexene carboxaldehyde sensitive-patients. Contact Dermatitis. 2009;61:152–62.

68. Garg S, Thyssen JP, Uter W, et al. Nickel allergy following European Union regulation in Denmark, Germany, Italy and the U.K. Br J Dermatol. 2013;169:854–8.

UV and Skin: Photocarcinogenesis

8

Allen S.W. Oak, Mohammad Athar, Nabiha Yusuf, and Craig A. Elmets

Abbreviations

6-4PP	Pyrimidine-pyrimidone 6-4 photoproduct
8-oxoG	8-Oxoguanine
AhR	Arylhydrocarbon receptor
AK	Actinic keratosis
AP	Apurinic or apyrimidinic
AP-1	Activator protein 1
APC	Antigen-presenting cell
ARNT	Aryl hydrocarbon receptor nuclear translocator
ATF	Activating transcription factor
BCC	Basal cell carcinoma
BER	Base excision repair
CHS	Contact hypersensitivity
CI	Confidence interval
CK1-α	Casein kinase 1 α
COX	Cyclooxygenase
CPD	Cyclobutane pyrimidine dimer
CS	Cockayne syndrome
DAMP	Damage associated molecular pattern
DFMO	α-Difluoromethylornithine
DHH	Desert Hedgehog
DNFB	Dinitrofluorobenzene
DTH	Delayed-type hypersensitivity
EGF	Epidermal growth factor
EGFR	Epidermal growth factor receptor
EMT	Epithelial-mesenchymal transition
ERK	Extracellular-signal-regulated kinase
FDA	Food and Drug Administration
FICZ	6-Formylindolo[3,2-b]carbazole
GG-NER	Global genome nucleotide excision repair
GLI	Glioma-associated oncogene
GSK3-β	Glycogen synthase kinase 3 β
GWAS	Genome-wide association study
HAF	Hyaluronic acid fragments
HB-EGF	Heparin-binding EGF
HCTZ	Hydrochlorothiazide
Hh	Hedgehog
IHH	Indian Hedgehog
IKK	IκB kinase
IL	Interleukin
IRR	Incidence rate ratio
IκB	Inhibitor of NF-κB
JNK	c-Jun amino-terminal kinase
LOX	Lipoxygenase
MAF	Musculoaponeurotic fibrosarcoma
MAPK	Mitogen-activated protein kinase
MCR1	Melanocortin 1 receptor
MM	Malignant melanoma
MMP	Matrix metalloproteinase
MyD88	Myeloid differentiation factor-88

A.S.W. Oak • M. Athar • N. Yusuf
C.A. Elmets, M.D. (✉)
Department of Dermatology, University of Alabama
at Birmingham, 1530 3rd Ave S, EFH 414,
Birmingham, AL 35294-0009, USA
e-mail: celmets@uab.edu

© Springer International Publishing Switzerland 2018
J. Krutmann, H.F. Merk (eds.), *Environment and Skin*,
https://doi.org/10.1007/978-3-319-43102-4_8

NAD	Nicotinamide adenine dinucleotide
NBCCS	Nevoid basal cell carcinoma syndrome
NEMO	NF-κB Essential modulator
NER	Nucleotide excision repair
NF-κB	Nuclear factor κ-light-chain-enhancer of activated B cells
NK	Natural killer
NMSC	Nonmelanoma skin cancer
NSAID	Nonsteroidal anti-inflammatory drug
ODC	Ornithine decarboxylase
ONTRAC	Oral Nicotinamide to Reduce Actinic Cancer
OR	Odds ratio
PAMP	Pathogen associated molecular pattern
PARP	Poly-adenosine diphosphate ribose polymerase
PI3K	Phosphoinositide 3-kinase
PKA	Protein kinase A
PUVA	Psoralen plus UVA
ROS	Reactive oxygen species
RRR	Relative rate reduction
SCC	Squamous cell carcinoma
SCUP-h	Skin Cancer Utrecht-Philadelphia-human
SCUP-m	Skin Cancer Utrecht-Philadelphia-murine
SHH	Sonic Hedgehog
SMO	Smoothened
SPF	Sun protection factor
SUFU	Suppressor of fused
TC-NER	Transcription-coupled nucleotide excision repair
TFIIH	Transcription factor IIH
TGF	Transforming growth factor
TLR	Toll-like receptor
TNCB	Trinitrochlorobenzene
TNF	Tumor necrosis factor
Treg	T regulatory cell, formerly known as suppressor T-cell
UV	Ultraviolet
UVA	Ultraviolet A (320–400 nm)
UVB	Ultraviolet B (280–320 nm)
UVC	Ultraviolet C (200–280 nm)
UVR	Ultraviolet radiation
VATTC	Veterans Affairs Topical Tretinoin Chemoprevention
VEGF	Vascular endothelial growth factor
XP	Xeroderma pigmentosum

8.1 Introduction

Basal cell and squamous cell carcinomas of the skin, grouped together under the umbrella term nonmelanoma skin cancers (NMSCs), are the most common malignancies in the human population. It is estimated that greater than five million NMSCs are treated in over three million persons in the United States annually [1]. This far exceeds the 1.7 million annual incidence of all other cancers combined [2]. In an era in which the incidence of most other cancers has either stabilized or decreased, the incidence of NMSC continues to increase. Since 1960, there has been a mean increase of 3–8% annually in NMSC for the Caucasian population of Australia, the USA, and Canada [3]. In addition, NMSCs are occurring in increasingly younger age groups [4, 5]. Fortunately, these malignancies have a low mortality rate and account for only 2000 deaths per year [6]. However, NMSCs can be locally destructive and are a source of considerable morbidity. This morbidity is accentuated by its propensity to develop in conspicuous locations in the body; 80% of NMSCs occur in the head and the neck. Potential sequelae include oral incompetence, nasal obstruction, facial nerve paralysis, and psychosocial distress [7]. The economic burden of skin cancer is considerable. The average annual total cost for skin cancer treatment increased from $3.6 billion to $8.1 billion between 2002–2006 and 2007–2011. Of the $8.1 billion, $4.8 billion is attributed to NMSC and $3.3 billion to melanoma [8].

Various modifiable and unmodifiable factors increase the risk of developing NMSC. These include skin type, a person's phenotypic traits that dictate his sun sensitivity; fair skin, blue eyes, red hair, and inability to tan have all been described to increase the risk of developing NMSC [3, 9]. As supportive evidence, the mean age-standardized annual incidence of NMSC in black, Asian, and white populations of South Africa between 2000 and 2004 are 9.3, 20, and 412.4 per 100,000 persons, respectively [10]. However, ultraviolet (UV) exposure is the most important modifiable risk factor for development of both squamous cell carcinoma (SCC) and basal cell carcinoma (BCC). This is supported by the observation that the incidence of NMSC increases with increasing proximity to the equator, regardless of gender or age [11]. Advanced age, immunosuppressive status, various inherited skin diseases (e.g., xeroderma pigmentosum), and male gender have all been linked to NMSC development [12].

The relationship between sunlight and skin cancer was first suggested by Enzière in the late nineteenth century, who noted that skin cancer of the lip was more common in poor countrymen with extensive outdoor sun exposure [13]. Unna observed that sailors, who were chronically exposed to large amounts of UV radiation (UVR) because of their occupation, were predisposed to developing SCC and BCC and they preferentially occurred in uncovered skin. Findlay provided the first experimental evidence to support a causal link between UVR and skin cancer when he produced malignant epitheliomas and papillomas in albino mice by chronically exposing them to UVR for 8 months or longer [14]. Based on Findlay's results, Roffo extended these observations by demonstrating that skin cancers could be produced in animal models by repeated exposures to natural sunlight [15].

Because of the clinical importance of the problem and the need to develop more effective methods for their prevention and therapy, there has been great interest in understanding the pathogenesis of UV-induced skin cancers. The discipline that investigates these issues is called photocarcinogenesis. There has been substantial progress in understanding how UVR causes skin cancer over the past several decades. New knowledge derived from experimental studies of photocarcinogenesis has contributed to fundamental principles of cancer biology, including, but not limited to, tumor oncogenes, tumor suppressor genes, the contribution of the immune system to control of cancer growth and development, molecular mechanisms of DNA damage, and methods of chemoprevention.

8.2 UV Radiation and NMSC

8.2.1 Basic Cancer Biology

Tumors form from the clonal expansion of cells with mutations in genes responsible for proliferation and differentiation. In the vast majority of cases, a single mutation is not sufficient to transform a normal cell into an autonomous neoplastic cell, which proliferates, grows out of control, becomes invasive, and metastasizes.

A gene may cause deleterious effects and contribute to carcinogenesis by producing too much or too little of its protein product. A gene that produces too much of its product is referred to as an oncogene (gain of function). An oncogene prior to the mutation is called a proto-oncogene. On the other hand, a tumor suppressor gene produces too little of its product (loss of function). An oncogene exerts a dominant effect on the cell, meaning that only a single copy of the defective gene is required to produce its effect. A tumor suppressor gene requires two defective copies, or two hits, and thus behaves in a recessive manner [16]. The recessive behavior displayed by tumor suppressor genes provides a conceptual basis for Knudson's "two-hit" hypothesis, which posits that patients with germline mutations in a tumor suppressor gene are born with the first hit already present and later acquire a second hit that results in tumorigenesis [17].

Carcinogens, agents capable of inducing tumors, are classified as genotoxic or non-genotoxic according to their ability to cause damage to the genome [18]. Both solar radiation and broad-spectrum UVR are genotoxic and non-genotoxic, and are classified as *known to be a human carcinogen* according to the most recently published *Report on Carcinogens* [19]. UVR is a relatively small, yet significant, part of solar radiation. Photons from infrared, visible, and UV spectra comprise the following percentages of terrestrial sunlight: 52, 45, and 3% [20]. Furthermore, UVR is artificially subdivided into UVA (320–400 nm), UVB (280–320 nm), and UVC (200–280 nm). When considering the quantity of UVR reaching the earth's surface, UVA is far more abundant (90–99%) than UVB (1–10%). However, UVB is more damaging per photon and accounts for about 90% of the carcinogenic dose of sunlight [18]. The highly mutagenic UVC is entirely filtered by the ozone layer. Both UVA and UVB are considered complete carcinogens, meaning that they are capable of driving all three stages of photocarcinogenesis on their own [21, 22].

8.2.2 The Multistage Model of Photocarcinogenesis

The development of UV-induced skin cancer proceeds through an orderly sequence of events in which molecular and biochemical changes accumulate in keratinocytes over long periods of time. Three separate stages have been delineated: initiation, promotion, and progression. During initiation, UV, largely UVB, induces DNA damage in keratinocytes, primarily in the form of cyclobutane pyrimidine dimers (CPDs) and pyrimidine-pyrimidone 6-4 photoproducts (6-4PPs). CPDs are formed when UV's photon energy opens up the double bonds between C-5 and C-6 of neighboring pyrimidines (cytosine or thymine), and two new covalent bonds are established between the pyrimidines. 6-4PP formation involves opening up a double bond between C-5 and C-6 of one of the neighboring pyrimidines, and establishing a new covalent bond between C-6 position of one and C-4 of the other. When unrepaired, DNA containing these lesions can be replicated and the resulting copy contains single or tandem base substitutions, which are almost always $C \rightarrow T$ transitions. Those CPDs and 6-4PPs, which cause mutations in tumor suppressor genes or oncogenes that control cellular proliferation, regulation, or differentiation, play a major role during the initiation stage of photocarcinogenesis. When UVB-induced CPDs and 6-4PPs occur in DNA,

there is a vigorous attempt to repair it through activation of DNA repair enzymes. When the lesions are not repaired by DNA repair pathways, the mutant cells fail to die and progress further. This is a rare event in healthy individuals, but chronic sun exposure increases its chance of occurrence. In such case, the initiated cell may gain a comparative advantage, such as resistance to apoptosis, and it continues proliferating during the subsequent promotion and progression stages. Furthermore, patients with xeroderma pigmentosum (XP), who are unable to repair UV-induced DNA damage, develop actinically damaged skin and NMSCs in sun-exposed areas as early as 3–5 years of age, compared to the mean age of 50–60 in the general population (Fig. 8.1) [23, 24].

During the second stage of photocarcinogenesis, termed promotion, repeated doses of UV induce chronic inflammation and encourage clonal expansion of initiated cells. Chronic inflammation is a predisposing factor for cancer promotion, as evidenced by an increased incidence of cancers in patients with chronic inflammatory conditions, such as reflux esophagitis, chronic hepatitis, chronic *H. pylori* infection, and inflammatory bowel disease [25]. The UV response, defined as the genes and signaling cascades activated by UVB, is a pseudo-growth response that mirrors the growth response triggered by mitogens. The UV response is initiated in both the nucleus via photoproduct formation

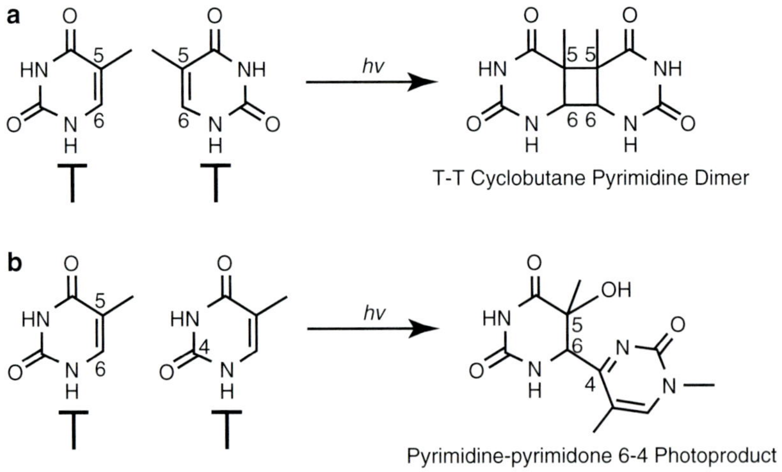

Fig. 8.1 UV-induced formation of a cyclobutane pyrimidine dimer (**a**) and a pyrimidine-pyrimidone 6-4 photoproduct (**b**)

and in the cytoplasm. UV-induced activation of signaling pathways in the cytoplasm induces activation of transcription factors: nuclear factor κ-light-chain-enhancer of activated B cells (NF-κB) and activator protein 1 (AP-1). The resulting changes in gene expression are responsible for chronic inflammation and increased proliferation. The end result is a premalignant actinic keratosis (AK, also known as solar keratosis), a reversible premalignant lesion. Five to twenty percentage of AKs progress into SCCs in 10–25 years in humans [26].

During progression, the third stage of photocarcinogenesis, with continued sun exposure, premalignant AKs undergo changes that grant genetic instability and lead to changes frequently observed in cancerous cells, such as acquired chromosomal aberrations. A key event during progression is epithelial-mesenchymal transition (EMT), in which epithelial cells undergo alterations in adhesion, cellular architecture, and morphology to remodel the extracellular matrix and increase their greater migratory capacity [27]. The induction of angiogenesis by proangiogenic cytokines, augmented production of cyclooxygenase-2 (COX-2), and increased synthesis of vascular endothelial growth factor (VEGF) are also essential during this stage. The end product of progression is an invasive squamous cell carcinoma.

UV-induced defects in immunological surveillance also contribute to the development of skin cancer. Immunological defense mechanisms have evolved that neutralize and/or eliminate malignant cells before they can become clinically apparent tumors. UV exposure inactivates these immunological defenses, thereby allowing neoplastic cells to grow and develop into invasive skin cancers.

Fortunately, progressing through all three stages of photocarcinogenesis usually takes years to decades. It is important to emphasize that while BCC and SCC both arise from keratinocytes and the vast majority are caused by UVR, they are quite different in their pathogenesis and behavior. BCCs are believed to develop de novo without a precursor lesion and can be locally invasive, but are almost never metastatic.

8.3 Photochemistry

Molecular organic photochemistry studies the ability of a photon, an elementary particle of electromagnetic radiation or light, to produce molecular and biochemical changes in target molecules. Furthermore, molecular organic photochemistry is composed of two fundamental branches: photophysics of organic compounds and photochemistry of organic compounds. Both photophysics and photochemistry require a photon as a driver of change. However, photophysics ultimately results in a net physical change, whereas photochemistry results in a net chemical change.

8.3.1 Fundamental Principles

Both photophysical and photochemical principles contribute to photocarcinogenesis. In the context of photocarcinogenesis, photophysics may be used to explain the formation of reactive oxygen species (ROS) and photochemistry may be used to explain CPD formation [28].

ROS are generated through incomplete reduction of molecular water to oxygen. UVR induces formation of various ROS: hydroxyl radical, hydrogen peroxide, singlet oxygen, and superoxide radical. Excessive ROS directly affect signal transduction and cause lipid peroxidation, both of which are deleterious and have been implicated in carcinogenesis. Physiologically, scavenger molecules and antioxidant enzymes attenuate the damaging effects of ROS [29].

A photon has the ability to break chemical bonds starting in the UV region of the electromagnetic spectrum. The photon energy required to ionize oxygen and hydrogen lies between 10 and 12 eV; thus, 10 eV is generally accepted as the lowest ionizing photon energy for any biological material [30]. Photon energies for UVA and UVB are 3.10–3.94 and 3.94–4.43 eV, respectively; since their photon energies do not exceed the photon energy required for ionization of oxygen and hydrogen, UVA and UVB are considered nonionizing [31]. Therefore, photocarcinogenesis is driven by the essentially nonionizing solar UV and is able to transverse the stratosphere.

8.3.2 Wavelength Dependence

Experimental data support the concept that within the UV spectrum, UVB wavelengths are the most carcinogenic. This was first recognized by Blum, who inserted plate glass between his UV light source and the animals that were being irradiated [32]. This procedure preferentially removes wavelengths within the UVB, but has a much smaller effect of UVA. Animals with the plate glass intervention developed significantly fewer tumors than those in which plate glass was not inserted. The wavelength dependence was further defined in action spectrum studies in which the narrower wavelength bands were evaluated for their relative efficacy [33]. Those studies showed that UVB wavelengths were much more effective than UVA in production of UV-induced tumors.

Studies by de Gruiji and colleagues were conducted in which hairless albino SKH:HR1 mice were exposed to chronic daily UV and the resulting carcinomas were serially observed and measured [34, 35]. The study statistically analyzed existing action spectra in the literature at the time, which all show effective tumor induction at

$\lambda < 300$ nm and a sharp decline in tumorigenic potential at $\lambda > 300$ nm. As a result, the SCUP-m (Skin Cancer Utrecht-Philadelphia murine) action spectrum was constructed. In terms of tumor induction, the SCUP-m action spectrum demonstrates maximal effectiveness at $\lambda = 293$ nm and a steep decline in tumorigenesis to 10^{-4} of maximum effectiveness at $\lambda = 340$ nm in the UVA region [35]. Action spectroscopy provides a method to help identify potential chromophores, the light absorbing molecules that initiate and drive the photobiological process of interest. The action spectrum for photocarcinogenesis corresponded well to the action spectrum for DNA, lending support to the hypothesis that the chromophore for photocarcinogenesis is DNA [36] (Fig. 8.2).

It should be noted that UVA without photosensitizers can also cause NMSCs in animal models, but it takes larger numbers of photons and a longer length of time to develop [37, 38]. In spite of this, there is growing concern of the role of UVA in skin cancer because of its potential role in melanoma [39], the widespread role of tanning beds which not only have an increased

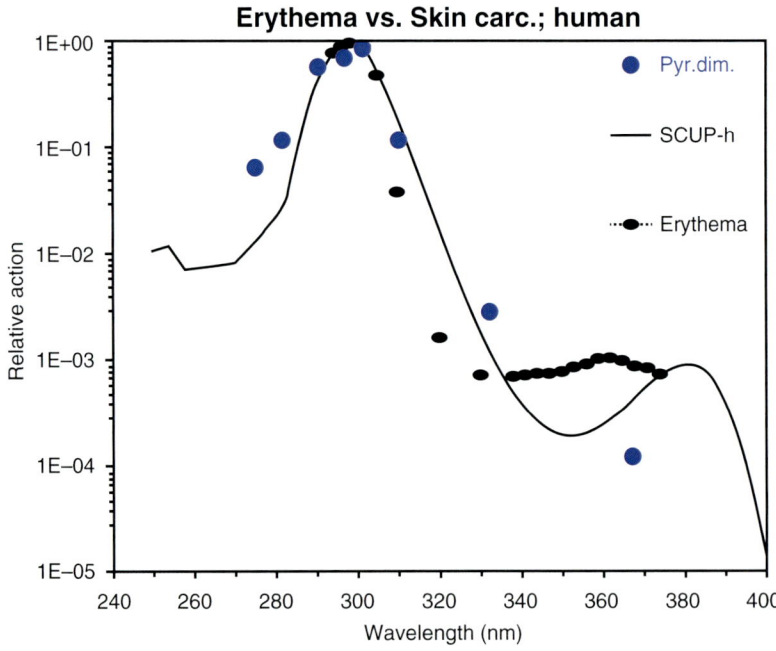

Fig. 8.2 Comparison of CPD formation in human skin and the action spectrum of photocarcinogenesis (Reprinted from J Photochem Photobiol B, 40, Black HS, deGruijl FR, Forbes PD, et al., Photocarcinogenesis: an overview, 29–47. Copyright (1997), with permission from Elsevier. The figure was generously provided by Dr. F.R. de Gruijl, Anders et al. 1995, De Gruijl et al. 1994)

risk for melanomas [40], but also SCCs and BCCs [41], and the chronic use of photosensitizing medications, which absorb UVA radiation and have been implicated in the predisposition to NMSCs.

8.3.3   Dose Response

Experimental data on chronic UV-exposed hairless mice support the notion that photocarcinogenesis is dose-dependent. Tumor induction time decreases with greater daily UV dose when examining hairless mice irradiated daily with UV. Only tumor initiation is shown to be dose-dependent while tumor growth is dose-independent [42].

8.3.4   Exposure Time

Sunlight is a known risk factor for both BCC and SCC, but exposure time is a critical factor that differs between the two. Intermittent sun exposure is a greater risk factor in BCC formation, whereas continuous sun exposure is a greater risk factor in SCC formation [43]. Furthermore, increased recreational sunlight exposure during childhood and teenage years is linked to a significantly higher risk of BCC than SCC [44].

8.3.5   Ozone Depletion

It is now widely accepted that there has been a dramatic drop in the amount of ozone present in the stratosphere. There was a 6% decline in the mid-levels of the Northern latitudes between 1980 and 1996, although there has been a 2% recovery between 1996 and 2009, due to measures taken to prevent the use of chlorofluorocarbons. Chlorofluorocarbons were widely used in aerosols. These compounds degrade the stratospheric ozone when released into the atmosphere. This has relevance to photocarcinogenesis because the ozone layer plays a significant role in diminishing the amount of UVR reaching the ter-

restrial surface. In the stratosphere, molecular oxygen (O_2) dissociates when irradiated with short-wavelength UV ($\lambda < 242$ nm) and forms ozone (O_3). Subsequently, ozone dissociates when it absorbs photons from UVR at wavelengths up to 320 nm; virtually all of UVC and most of UVB are absorbed in this manner prior to reaching the earth's surface [45].

Many believe that depletion of the stratospheric ozone layer can partially account for the rising incidence of NMSC. Mathematical models predict that there will be a 2% increase in the incidence of NMSC for every 1% decrease in stratospheric ozone [46]. More recently, a regression analysis between measured solar UVB and increased skin cancer incidence in Korea suggests that the amplification factor is even greater. For every 1% decline in the stratospheric ozone, there is a 2.98–3.70% increase in BCC and SCC [47].

Fortunately, controlled measures have been taken to mitigate the damage to the ozone layer since its destruction was first brought to international attention. The Montreal Protocol, an international treaty declared in 1987 and amended several times since then, has played a major role in curbing the damage to stratospheric ozone. A full implementation of the Montreal Protocol is predicted to prevent approximately 1.6 million skin cancer deaths and 280 million cases of skin cancer for Americans born between 1890 and 2100 [48].

8.3.6   Indoor Tanning as a Risk Factor for Photocarcinogenesis

Indoor tanning is a popular way to obtain an artificial tan for cosmetic purposes. The fluorescent lamps used by tanning parlors contain up to ten times the amount of UVA in sunlight. Furthermore, the output of modern sunbeds equipped with low-pressure fluorescent lamps is not entirely UVB-free since the lamps always emit some UVB, about 1–2% of the spectrum [49]. Indoor tanning is a known modifiable risk factor for NMSC and melanoma. A meta-analysis of ten studies reveals relative risks of BCC and

SCC for those exposed to indoor tanning versus those never exposed to be 1.29 (95% confidence interval (CI) = 1.08–1.53) and 1.67 (95% CI = 1.29–2.17), respectively [41]. The odds ratio for melanoma development for those ever using an indoor tanning bed is 1.16 (95% CI = 1.05–1.28) based on a meta-analysis of 31 studies in North America, Europe, and Oceania [40]. The public health concern regarding indoor tanning is exacerbated by its prevalence, especially in the younger population. The prevalence of ever exposure of indoor tanning is 35.7% (95% CI = 27.5–44.0%) for adults and 55.0% (95% CI = 33.0–77.1%) for university students. The same study attributed 450,000 NMSC cases and greater than 10,000 melanoma cases each year to indoor tanning in the USA, Europe, and Australia [50].

8.4 UV-Induced Photodamage to DNA and Repair Mechanisms

8.4.1 UV-Induced DNA Photoproducts

DNA is exceptionally vulnerable to UV-induced damage. Pyrimidines are approximately ten times more UV-sensitive than purines [51]. In 1958, Beukers and Berends reported that an irradiation product is formed after UV-irradiating a frozen solution of thymine with a low-pressure mercury lamp. NMR, infrared, and crystallographic data revealed that the photoproduct is a CPD [52].

CPDs, arguably the most important lesions during initiation, are 3–4 times more abundant than 6-4PPs and contribute to about 80% of UVB-induced mutations in mammalian cells

[53]. 6-4PPs, which induce a larger distortion and are repaired faster than CPDs, may photoisomerize to Dewar valence isomers upon further irradiation with wavelengths greater than 280–290 nm. Dewar isomers are considered less mutagenic and play a less important role [54]. Both CPDs and 6-4PPs, if unrepaired, may cause DNA polymerase to misread the template strand and cause a mutation. In the case of CPDs, mutations occur in a homogenous and identifiable manner frequently found in skin tumors and hence are named UV signature mutations. UV signature mutations, C → T or CC → TT mutations at dipyrimidine sites, suggest past UV exposure and serve as useful experimental markers. Signature mutations form instantaneously and appear structurally indistinguishable as part of the newly synthesized DNA to the DNA repair enzymes. In other words, initiation is irreversible and may occur at any time during a person's life (Fig. 8.3).

UVC, UVB, UVA, and visible light are able to reach the epidermis, papillary dermis, reticular dermis, and subcutis, respectively [55]. The maximal absorption of DNA in vitro occurs at $\lambda = 260$ nm with significant absorption in the UVB range [51]. The induction of CPD formation occurs at a vanishingly small, yet quantifiable, rate at $\lambda > 300$ nm. Therefore, the λ_{max} of 293 nm of SCUP-m, which falls between 260 and 300 nm, occurs at a range of UV in which photons are effectively absorbed by DNA and able to induce CPD formation. The λ_{max} of SCUP-m directly matches the wavelength accounting for the maximal rate of epidermal pyrimidine dimer action spectrum in hairless mice [56]. De Gruijl and van der Leun have extrapolated a human action spectrum for CPDs, accordingly named SCUP-h (Skin Cancer Utrecht-Philadelphia-

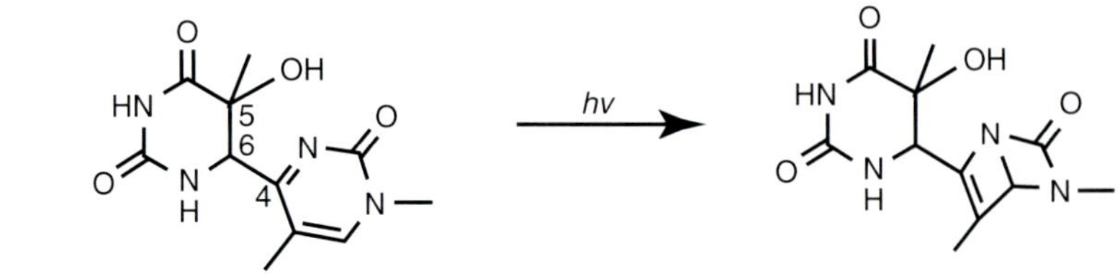

Fig. 8.3 Photoisomerization of a pyrimidine-pyrimidone 6-4 photoproduct into its Dewar valence isomer

Pyrimidine-pyrimidone 6-4 photoproduct   Dewar valence isomer

human), from the SCUP-m action spectrum by accounting for differences between mouse epidermis and human epidermis [57]. SCUP-h's λ_{max} of 299 nm closely matches the action spectrum of thymine dimer formation in humans [58].

Formation of CPDs was almost wholly attributed to UVB until recently. The mode of UVA damage was primarily thought to be due to oxidative damage. ROS react with guanine and cause the formation of oxidation products, such as 8-oxo- or 8-hydroxy-deoxyguanosine adducts, that result in a G → T or T → G transversion [59]. Mouret et al. have demonstrated that CPDs, but not 6-4PPS, are found at a higher yield than 8-oxoguanine (8-oxoG), the most commonly found oxidatively generated DNA lesion, in whole human explant skin and cultured cells irradiated with UVA [60]. In doing so, they have concluded that CPD is the predominant lesion in UVA-irradiated skin. Tewari et al. confirm the finding in healthy human volunteers in vivo [61]. The ability of UVA to directly damage DNA, even at a far less effective rate than UVB, has important implications due to UVA's relative abundance in sunlight and its use in both phototherapy and recreational indoor tanning (Fig. 8.4).

8.4.2 Repair Mechanisms

Unrepaired CPDs and 6-4PPs distort the helical structure of DNA and hinder both replication and transcription. Fortunately, only a small portion of UV-induced photoproducts causes mutations since repair and bypass mechanisms both exist. In humans, these DNA lesions are repaired primarily by excision repair, which replaces damaged DNA with new undamaged nucleotides. There are two major categories of excision repair:

base excision repair (BER) and nucleotide excision repair (NER). Bulky CPDs and 6-4PPs can only be repaired endogenously by NER in humans, but BER can repair other non-bulky oxidative DNA modifications, such as 8-oxoG. In BER, DNA glycosylase recognizes a single defective base, cleaves it, and leaves an apurinic or apyrimidinic (AP) site. AP endonucleases or AP lyases, which, respectively, nick the DNA at the 5′ or 3′ site relative to the AP sites, subsequently remove the AP sites. Phosphodiesterase removes the deoxyribose phosphate residue. In a process called unscheduled DNA synthesis, DNA polymerase β then fills the gap and DNA ligase seals the gap around the newly synthesized oligonucleotides [62, 63]. The ability of BER to repair different types of damaged bases is dictated by the substrate specificity of DNA glycosylase and many different glycosylases exist [64]. For example, 8-oxoG DNA glycosylase 1 recognizes oxidized guanine bases in humans and its loss causes cellular hypersensitivity to UVA.

NER involves similar basic steps of recognition: DNA incision, DNA synthesis, and ligation. Two distinct branches of NER exist, marked by different modes of DNA damage recognition, which are transcription-coupled repair (TC-NER) and global genome repair (GG-NER). TC-NER, the sole method of CPD removal in mouse epidermis, preferentially repairs the transcribed strand of an active gene and GG-NER repairs all other parts of the genome [65]. TC-NER is initiated when RNA polymerase stalls at the site of a CPD or 6-4PP. Subsequently, CSA and CSB (protein products of Cockayne syndrome complementation group A and group B genes) are recruited to the site. On the other hand, GG-NER is initiated when XPC and XPE (protein products of xeroderma pigmentosum complementation

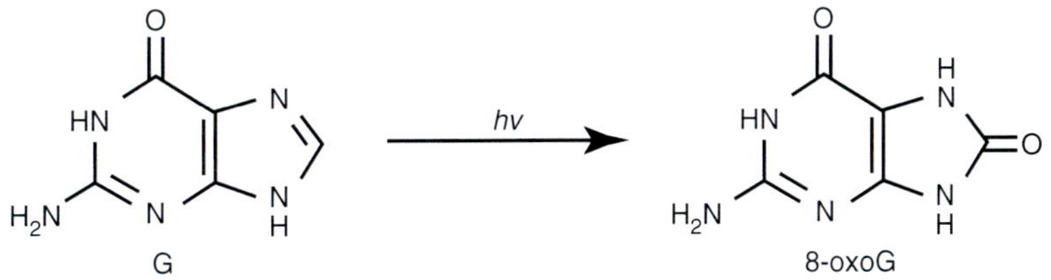

Fig. 8.4 Formation of 8-oxoG

group C and E genes) bind the DNA strand containing the photoproduct. Subsequent steps are identical for both TC-NER and GG-NER. First, replication protein A and transcription factor IIH (TFIIH) form an open complex around the UV-damaged site. Two subunits of TFIIH, XPD and XPB, then act as a DNA helicase and unwind the DNA. XPG and XPF, respectively, incise the 3′ and the 5′ ends of the damaged strand. The damaged portion of DNA is released, the gap is filled in by DNA polymerase δ or ε, and the repair segment is sealed by DNA ligase. NER is not crucial for viability in utero, but a defect in one of the involved proteins can result in XP, a disease characterized by defective DNA repair and increased susceptibility to UV-induced NMSC and melanoma as early as 3–5 years of age.

XP is an autosomal recessive disease, which may occur from germline mutations in any one of its seven different complementation groups, corresponding to defects in XPA to XPG. Thus, XP patients have a defect in GG-NER. On the other hand, other diseases with inherited NER defects, such as Cockayne syndrome (CS), UV-sensitive syndrome, and trichothiodystrophy, consistently yield a clinical phenotype of photosensitivity, but without an increased risk of malignancy. NER defects have other clinical manifestations, such as neurologic impairment and significantly reduced life span [66].

8.4.3 Bypass Mechanism

Not all DNA photoproducts are repaired, and a bypass mechanism is present for such situations. When DNA polymerase encounters an unrepaired photoproduct during replication, it usually stalls and detaches. In this situation, a DNA polymerase capable of bypassing the lesion is recruited to continue replication, a process called translesion synthesis. DNA polymerase η is one such enzyme and it is recruited to CPD-containing sites. DNA polymerase η acts as a "molecular splint" to stabilize the CPD-containing site to ensure the insertion of correct complementary nucleotides, and detaches thereafter. Patients with a deficiency in DNA polymerase η exhibit a phenotype called XP variant that is indistinguishable from that of

XP. DNA polymerase η may also explain why CPDs are most commonly generated at T-T sites while a signature mutation arises from sites containing cytosine instead. One proposed hypothesis posits that the unstable cytosine or methylcytosine in CPDs subsequently deaminates to uracil or thymine, respectively, which in turn directs an insertion of adenine on the newly synthesized complementary strand by DNA polymerase η during subsequent replication [67–71]. In addition, CPDs accelerate the cytosine deamination rate a millionfold [72]. This finding is consistent with the proposed "A" rule, which states that DNA polymerase inserts adenines by default in DNA sites it cannot interpret. Thus, a DNA polymerase will insert two adenines on the newly synthesized complementary strand when attempting to replicate a site containing a C-C CPD. On the next round of replication, two thymines are inserted on the site that previously contained two cytosines (CC → TT) (illustrated in Fig. 8.7). DNA polymerase η has been shown to accurately and efficiently bypass T-T CPDs by inserting adenines in yeast and in humans. DNA polymerase η also faithfully replicates and incorporates adenines into the complementary strand when replicating T-U-CPDs, which would occur as a result of deamination from T-C CPD [68, 73, 74].

8.4.4 Photosensitizing Agents

Photosensitizing drugs can, in certain instances, increase the risk of skin cancer. Psoralen plus UVA (PUVA) photochemotherapy, used to treat psoriasis, vitiligo, cutaneous T-cell lymphoma, and other dermatological diseases, relies on drug-induced photosensitivity as part of its therapeutic mechanism. Long-term PUVA therapy has been linked to photocarcinogenesis [75]. The PUVA Follow-Up Study, a prospective study that followed the cohort of patients first treated with PUVA in 1975–1976, concluded in 2005. Of the 759 patients who survived to the end of the trial, 37% developed one or more NMSCs. In the cohort, the incidence of SCCs occurred was approximately 30 times greater than the expected incidence in the general population, while BCCs occurred about 5 times more frequently. The increase in NMSC incidence was proportional to the number of treatments given, and the incidence rate ratio increased far more steeply for SCC than BCC [75].

The use of other photosensitizing drugs has been shown to increase the chance of NMSC as well. The corresponding rates of increase for BCC and SCC attributed to photosensitizing drugs seem to differ significantly, and the difference is not uniform for different drug classes. The New Hampshire Skin Cancer Study, one of the few population-based NMSC studies in the world, reported odds ratios (OR) for SCC, BCC, and early-onset BCC (diagnosed prior or at the age of 50) related to ever use of photosensitizing medication as 1.2 (95% CI = 1.0–1.4), 1.2 (95% CI = 0.9–1.5), and 1.5 (95% CI = 1.1–2.1), respectively. Interestingly, the OR of BCC (1.9, 95% CI = 1.3–2.8) was higher than that of SCC (1.4, 95% CI = 0.9–2.1) for photosensitizing antimicrobial use. The OR was even higher for early-onset BCC (OR 2.0, 95% CI = 1.3–3.4). The same relationship was found when looking only at tetracycline use [76, 77]. On the other hand, certain antihypertensive drugs have been linked to increased risk of SCC, and possibly to BCC and malignant melanoma (MM). The study of the Danish Cancer Registry by Jensen et al. revealed that use of photosensitizing diuretics were a risk factor for SCC (incidence rate ratio (IRR) 1.21, 95% CI = 1.04–1.40) and MM (IRR 1.19, 95% CI = 1.01–1.41), but not BCC (IRR 0.96, 95% CI = 0.90–1.03). Of the diuretics studied, amiloride, hydrochlorothiazide (HCTZ), and the combination of both yield the highest incidence risk ratio for SCC. Combination therapy of amiloride and HCTZ for more than 5 years demonstrate an IRR of 1.97 (95% CI = 1.49–2.62) [78]. In another study of the Danish Cancer Registry, Schmidt et al. confirmed the link between ever use of diuretic and increased risk of SCC (OR 1.19, 95% CI = 1.06–1.33), but no link between diuretic use and MM or BCC was found regardless of duration. Again, the highest risk of SCC was observed in combination therapy of potassium-sparing agent and a low-ceiling diuretic, such as a thiazide (OR 2.68, 95% CI = 2.24–3.21) [79]. Finally, long-term daily use of most photosensitizing drugs does not seem to increase the risk of skin cancer, barring some exceptions. A population-based cohort study of 4.7 million Danish residents assessed the risk of skin cancer with a long-term daily use of 19 known photosensitizing agents. Only furosemide and methyldopa were linked with $a \geq 20\%$ increased risk of SCC or BCC [80]. Voriconazole, used to treat systemic fungal infections has been linked to aggressive and even metastatic squamous cell carcinomas [81–86]; there are case reports of melanomas occurring in individuals taking this medication as well [87].

8.5 Oncogenes and Tumor Suppressor Genes

The causative role of UV-induced DNA damage in photocarcinogenesis is highlighted by the overwhelming prevalence of signature mutations in oncogenes and tumor suppressor genes that are associated with NMSC.

8.6 TP53

TP53 is a tumor suppressor gene that is mutated in over 90% of SCCs. *TP53* encodes p53, a protein that detects DNA damage and triggers cell cycle arrest or apoptosis. A mutation in one copy of *TP53* confers partial deficiency in sunlight-induced apoptosis, but cells with two mutated *TP53* alleles lose a checkpoint that is critical in maintaining genome integrity. Subsequently, they become aneuploid [88, 89]. Sequencing data from SCCs indicate that about 70% of *TP53* mutations are C → T mutations and approximately 10% are CC → TT mutations [88]. This mutation pattern is remarkably similar to that found in AKs [90]. The presence of p53 mutations in AKs, combined with evidence that as UVB irradiation increases, the cells with aberrant p53 immunostaining increases, suggests that p53 mutations happen relatively early during photocarcinogenesis [91]. However, two functional copies of *TP53* need to be inactivated prior to SCC formation; in a typical SCC, a signature mutation is found in one copy of *TP53* while the other copy is deleted. Sequencing the p53 coding regions of BCCs reveals a point mutation in 56% of BCCs, 100% of which occur at adjacent pyrimidine sites [72]. In contrast, signature mutations are not found in *TP53* mutations in cancers of internal organs [88, 92]. In fact, studies of p53-negative homozygous mice reveal that UVR is a mandatory component for cutaneous SCCs. p53-null mice spontaneously develop various tumors, primarily lymphomas and sarcomas, but only develop primary cutaneous tumors, including SCCs, with chronic UV irradiation [93, 94].

8.7   PTCH

UV-induced mutations in the tumor suppressor gene *TP53* is crucial for both SCC and BCC pathogenesis. However, *PTCH*, a tumor suppressor gene and the human homolog of the *Drosophila melanogaster patched*, provides a distinguishing point between BCC and SCC carcinogenesis since mutations in the *PTCH* genes play a much more important role in BCC carcinogenesis. PTCH, the protein product of *PTCH*, is part of the Hedgehog signaling pathway, which determines anterior-posterior relationships (segment polarity) and contributes to neural tube patterning during embryogenesis. The mammalian Hedgehog (Hh) signaling involves three Hh ligands, Sonic Hedgehog (SHH), Desert Hedgehog (DHH), and Indian Hedgehog (IHH), that bind a negative regulatory receptor, PTCH. PTCH normally binds and inhibits Smoothened (SMO), a transmembrane G-protein receptor. Without an Hh ligand, PTCH represses SMO activity and prevents its ciliary accumulation. When bound by an Hh ligand, PTCH displaces from the cilia and SMO accumulates and activates in the cilia. SMO then triggers a signaling cascade that activates glioma-associated oncogene (GLI) transcription factors, GLI1, GLI2, and GLI3. Activated GLIs, the final effectors of the Hh signaling pathway, are capable of inducing expression of genes such as *MYC*, *SNAIL,* and *BCL2* that reg-

ulate survival, differentiation, and proliferation [95]. Furthermore, Hh signaling is essential for maintaining the stem cell population in the skin, regulating the hair cycle, and controlling sebaceous gland development [66, 96]. Hedgehog signaling pathway's role in BCC carcinogenesis has been well documented and many of its members have been studied extensively as potential drug targets to treat advanced BCC in adult patients who are not candidates for surgical resection or radiation. Small molecule SMO receptor antagonists, vismodegib and sonidegib, were recently approved in the USA for treatment of BCC [97, 98].

Much of the work elucidating the role of *PTCH* in BCC pathogenesis has come from studying patients with nevoid basal cell carcinoma syndrome (NBCCS), also known as Gorlin syndrome. Gorlin syndrome is an autosomal dominant disorder with an estimated prevalence of 1 per 56,000 and is characterized by multiple early BCCs, jaw keratocysts, medulloblastomas, progressive intracranial calcification, and dyskeratotic pitting of hands and feet. Other developmental defects, such as skeletal and midline brain malformations, are also common [99, 100]. Over their lifetime, patients with Gorlin syndrome develop dozens to thousands of BCCs. Hahn et al. mapped *PTCH* to a precise location of 9q22.3 and revealed germline and somatic *PTCH* mutations in those with Gorlin syndrome and sporadic BCC, respectively [101]. Subsequent studies have identified mutations in at least one allele of *PTCH* in >85% of sporadic BCCs [102]. 10% have activating mutations in *SMO*, which keep the Hh signaling pathway constitutively active [103]. *PTCH* mutation spectra reveal C → T or CC → TT transitions at dipyrimidine sites in 68% of exonic mutations and 82% of intronic mutations. No correlation was found between clinical phenotype, age of first BCC or number of BCCs, and proportion of UV-associated mutations [104]. The importance of *PTCH* in photocarcinogenesis is reinforced in studying XP patients, in which 80–90% of BCCs from XP patients show a mutation in *PTCH*, of which about 80% are UV specific. A high level of UV

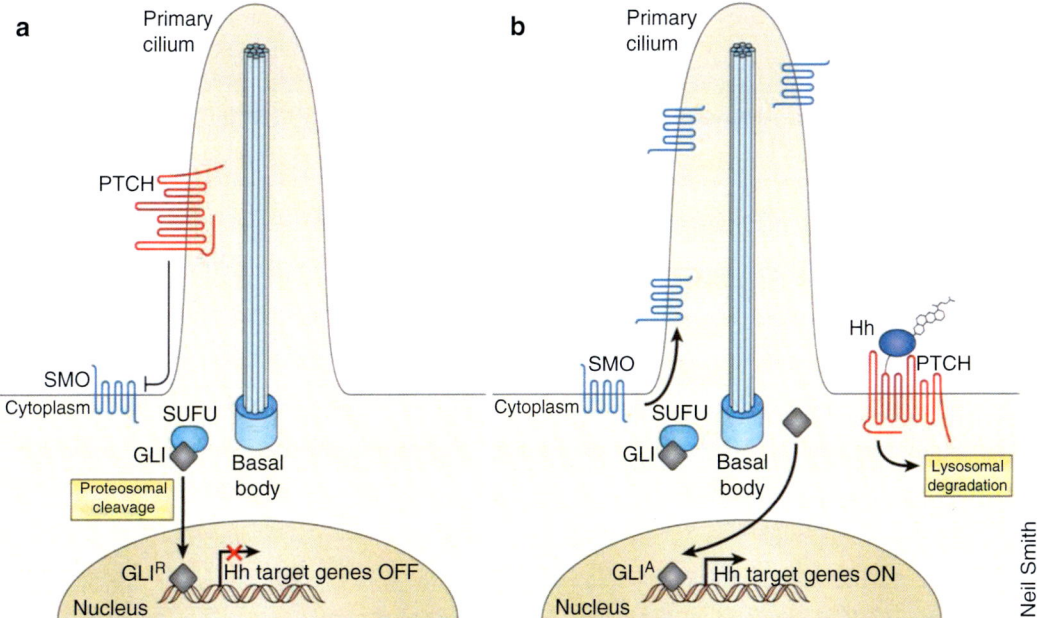

Fig. 8.5 The mammalian Hh signaling pathway: key components and signal transduction. (**a**) In the absence of Hh ligand, PTCH localizes in the cilia and represses SMO activity by preventing its trafficking and localization to the cilia. GLI transcription factors are sequestered in the cytoplasm by several protein mediators, including protein kinase A (PKA), glycogen synthase kinase 3 β (GSK3-β), casein kinase 1 α (CK1-α), and suppressor of fused (SUFU). GLI undergoes proteasomal cleavage and the resulting repressor form (GLI^R) translocates to the nucleus and inhibits translation of Hh target genes. (**b**) On ligand binding, PTCH is displaced from the cilia, thereby allowing ciliary accumulation and activation of SMO. Activated SMO orchestrates a signaling cascade that eventually results in translocation of an activated form of GLI (GLI^A) to the nucleus, where it induces expression of Hh target genes (Reprinted by permission from Macmillan Publishers Ltd.: [Nature Medicine] [95], copyright (2013))

specific mutations is found in BCCs from XP patients with *SMO* mutations [105].

The first animal model of BCC was developed by Oro and colleagues by fusing *Shh* to the promotor of keratin 14 (K14). The K14 *Shh*-induced cutaneous BCCs appear within 4 days of skin development [106]. Subsequently, additional transgenic mice have been generated which target downstream members of the pathway, using a K5 promoter to drive the expression of SMO-M2, a constitutively active SMO mutant, or *Gli1* or *Gli2* [107–112]. All successfully yielded spontaneous development of BCC and have led to the conclusion that continued Hh signaling pathway is required for BCC survival and proliferation. Finally, Aszterbaum and colleagues [113] were the first to develop a successful mouse model of UV-induced BCC tumorigenesis. Using *PTCH*$^{+/-}$ mice, they were able to induce the formation of BCCs and trichoblastoma-like tumors with chronic exposure to either UV or ionizing radiation (Fig. 8.5).

8.8   CDKN2A

Other regulators of the cell cycle have been implicated in photocarcinogenesis. *CDKN2A* encodes p16^{INK4a} and p14^{ARF}, both of which inhibit cell cycle progression at the G$_1$/S checkpoint. *CDKN2A* inactivation by promoter methylation has been implicated in SCC pathogenesis. Inactivation of *CDKN2A* was found in 5 (4 SCCs and 1 AK) out of 21 tumors analyzed (24%) [114]. In addition, mutation analysis of *CDKN2A* in Greek patients found mutations in 9% of SCCs, but not in BCC or Bowen's disease samples [115].

8.9   ras

Mutations in *ras* oncogenes, *Harvey (Ha)*, *Kirsten (Ki)*, *and N-ras*, which are monomeric GTPases, have been extensively studied since the 1990s, but their involvement in NMSCs is controversial [16]. Mutated *ras*, the first identified human oncogene, is implicated in about 20% of human cancers and therefore was an attractive potential target gene for NMSC around the time it was discovered. In humans, activating point mutations in *ras* almost always occur on codons 12, 31, and 61. Although a mutation frequency of *ras* in 10–20% of NMSCs in humans and mice has been suggested, the prevalence of *ras* mutations ranges from 10 to 40% in different studies [116, 117].

8.10   Findings from Genome-Wide Association Studies

Rapid advances in DNA sequencing in the past two decades have made it possible to examine disease-associated genetic variants for a large number of individuals. Genome-wide association studies (GWAS) in dermatology have helped identify susceptibility variants in the genome, which may aid risk stratification for skin cancer in the future. Genes involved in the pigmentation pathway make readily apparent targets for risk stratification since basal pigmentation is clearly linked to UV susceptibility. Evidence that basal pigmentation has a photoprotective effect is supported by a tenfold difference in the incidence of skin cancer between Caucasians and African Americans in the USA. Melanocortin 1 receptor (*MCR1*) variants, which are associated with lighter skin, red hair, and poor tanning response, have been linked to an increased risk of NMSC by a factor of 3.2 and cutaneous malignant melanoma by a factor of 2 [118]. Variants in *ASIP* (encoding agouti signaling protein, which antagonizes the interaction between α-melanocyte stimulating hormone and MCR1), and *TYR* (encoding tyrosinase, rate-limiting enzyme of the melanin synthesis pathway) have been linked to an increased risk of both cutaneous malignant melanoma and BCC [119]. Certain variants, such as L374F in *SLC45A2* (previously

MATP, which encodes a transporter protein which mediates melanin synthesis), show an impressive association with BCC (OR 1.97, 95% CI = 1.63–2.38), SCC (OR 2.71, 95% CI = 1.88–3.92) and cutaneous malignant melanoma (OR 2.95, 95% CI = 2.42–3.60). Interestingly, the same study identified a susceptibility variant in *KRT5*, which encodes keratin 5 (K5) that confers susceptibility to BCC (OR 1.35, 95% CI = 1.23–1.50) and SCC (OR 1.25, 95% CI = 1.05–1.48) [120].

8.11   Inflammatory Responses

Repeated doses of UV establish and maintain a chronic inflammatory environment that produces a selective proliferative advantage in mutant keratinocytes over non-mutated cells. Inflammatory changes are brought about by UV-induced generation of cytokines, free radicals and arachidonic acid metabolites, as well as the activation of signaling pathways that in turn affect cellular proliferation, survival, and differentiation.

8.12   Chromophore for the Nonnuclear UV-Induced Signaling Cascades

Recently, tryptophan has been proposed as a chromophore for the nonnuclear UVB-induced signaling cascades with cytoplasmic binding of its derivative, 6-formylindolo[3,2-*b*]carbazole (FICZ), to arylhydrocarbon receptor (AhR) acting as an initiating step. AhR is a cytosolic receptor that also mediates toxicity of carcinogenic polycyclic aromatic hydrocarbons. AhR activation in the cytosol relays the signal in two directions: toward the nucleus and toward the cell membrane. AhR activation relays the signal toward the nucleus as follows. Upon binding its ligand, AhR sheds its associated proteins, Hsp90 and c-src (pp60[src]) and translocates to the nucleus to form a heterodimer with the aryl hydrocarbon receptor nuclear translocator (ARNT). In the nucleus, the AhR/ARNT heterodimer alters gene transcription by activating genes, such as cyto-

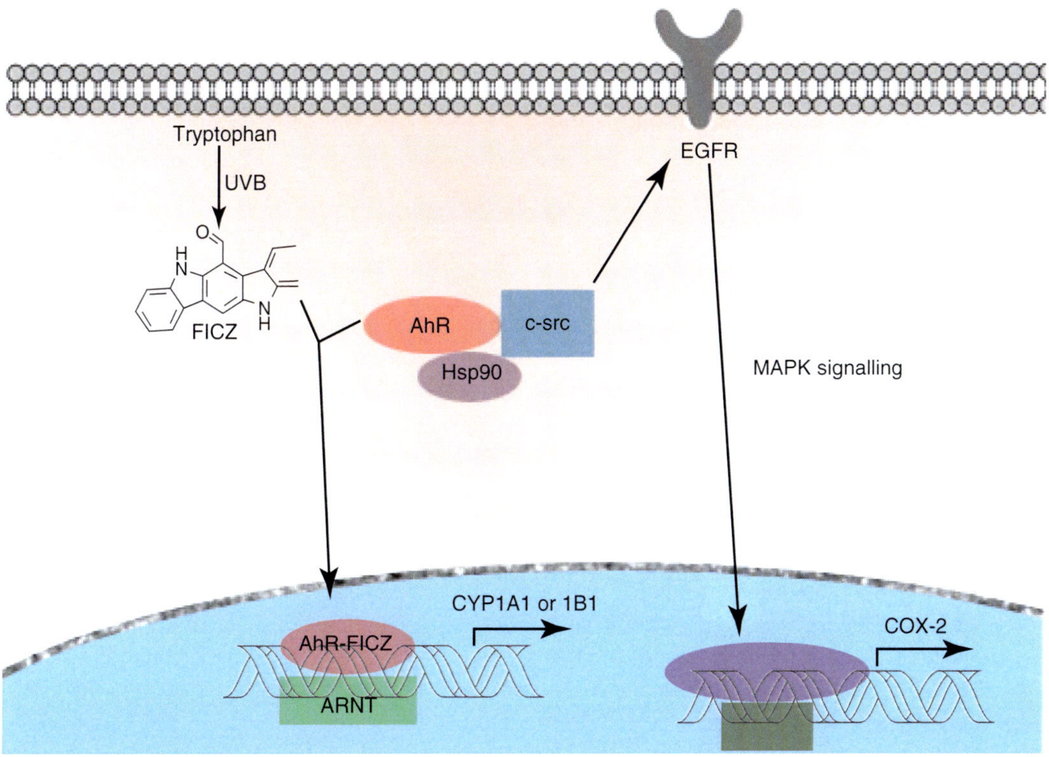

Fig. 8.6 In the cytoplasm, UVB-induced tryptophan photoproduct, 6-formylindolo[3,2-*b*]carbazole (FICZ), binds and activates arylhydrocarbon receptor (AhR). AhR activation starts a bidirectional signaling cascade. After shedding its associated proteins, Hsp90 and c-src, AhR translocates to the nucleus to dimerize with the aryl hydrocarbon receptor nuclear translocator (ARNT) where they transcribe xenobiotic response elements through activation of genes, such as cytochrome P450 1A1 or 1B1. After dissociating from the AhR/c-src/Hsp90 complex, c-src kinase phosphorylates EGFR and the signal is relayed downstream via a MAPK signaling pathway; COX-2 is a known downstream target

chrome P450 1A1 or 1B1 [121]. On the other hand, AhR activation relays the signal toward the cell membrane since the protein tyrosine kinase, c-src, is thought to activate epidermal growth factor receptor (EGFR), thus triggering the mitogen-activated protein kinase (MAPK) signaling pathway. COX-2 is a known downstream target [122–125]. Using HaCaT cells and C57BL/6 mice, Fritsche and colleagues were the first to describe UVB-induced AhR activation by demonstrating: (a) intracellular formation of FICZ in vivo upon UVB irradiation, (b) UVB-triggered AhR translocation to the nucleus and subsequent transcription induction of an AhR-dependent genes, (c) EGFR internalization and phosphorylation of EGFR-dependent ERK1/2, and (d) failure to demonstrate UVB-induced AhR translocation to the nucleus, transcription induc-

tion of AhR-dependent genes, and EGFR internalization in AhR-KO cells [126]. Interestingly, this study did not demonstrate AhR translocation when the HaCaT cells were irradiated with UVA, although the irradiation dose (30 J/cm^2) was comparable to that (250 kJ/m^2=25 J/cm^2) used by Vile and colleagues in their study of NF-κB activation. Thus, all signaling pathway activation cannot be explained by AhR, according to what is currently known about the pathway (Fig. 8.6).

8.13   NF-κB and the NF-κB Signaling Pathway

The NF-κB signaling pathway, named after its core member NF-κB, lies at the heart of the inflammatory network. In most unstimulated cells, the inhibitor of NF-κB (IκB) forms a dimer with NF-κB, thus storing NF-κB in the cytoplasm as an inactive complex, NF-κB-IκB. In mammals, 5 NF-κB proteins, RelA (p65), RelB, c-Rel, NF-κB1 (p105/p50), and NF-κB2 (p100/52), form homodimers or heterodimers among themselves to turn on their own characteristic genes. Once activated, NF-κB regulates production of almost all gene products linked to inflammation (IL-1, IL-6, COX-2, and 5-lipoxygenase (5-LOX)). Furthermore, NF-κB is constitutively active in most tumor cells since its gene products are anti-apoptotic and pro-proliferative [127]. It is also linked with progression of AKs to invasive NMSCs since the genes that are activated by NF-κB also aid invasion, metastasis, and angiogenesis (VEGF, adhesion molecules, TWIST, CXCR4, matrix metalloproteinases (MMP)).

UV induces the NF-κB signaling pathway and it does so without requiring cellular detection of DNA damage. A UV-mediated signaling pathway that does not originate from detecting nuclear DNA damage was first identified by Devary and colleagues, who showed that NF-κB and AP-1 induction by UVC does not involve nuclear signal generation using enucleated HeLa cells. Furthermore, they posited that the initiating event that triggers NF-κB occurs at or near the plasma membrane [128]. Moreover, cytosolic extracts from the A431 human epidermoid carcinoma cell line were exposed to 200 J/m² UVB, a dose at which 90% of cells are viable. Dissociation of preexisting NF-κB-IκB, an inactive complex, was detected in cytosolic extracts free of chromosomal DNA following UVB irradiation in a dose-dependent manner. Extraction of membrane components by centrifugation markedly diminished NF-κB activity [129]. NF-κB activation by a nonlethal dose of UVA (250 kJ/m²) was then demonstrated in enucleated cultured human skin fibroblasts, thereby demonstrating that UVA-dependent NF-κB activation is correlated with oxidative membrane damage [130]. The series of enucleation and cytosolic extract experiments thus demonstrate that UV induction NF-κB signaling pathway, at wavelengths and doses pertinent to photocarcinogenesis, (a) occurs independently from chromosomal DNA damage and (b) is initiated at or near the plasma membrane.

Physiologically, the NF-κB signaling pathway is activated by three distinct types of receptors (tumor necrosis factor (TNF)-α receptor, IL-1 receptor, and Toll-like receptors) which share a similar mechanism of NF-κB activation. Toll-like receptors (TLRs), which are homologs of Toll receptors in *Drosophila,* are members of the innate immune system that influence acquired immune responses. Most TLRs are coupled to a cytoplasmic signaling cascade that includes the adaptor protein myeloid differentiation factor-88 (MyD88), which ultimately leads to transcription of genes controlling inflammatory responses. TLRs are present on macrophages, dendritic cells, and other cells that recognize exogenous pathogen-associated molecular patterns (PAMPs), endogenous damage-associated molecular patterns (DAMPs), and microorganism-associated molecular patterns (MAMPs) that come from pathogens and commensal bacteria [131]. Different TLRs recognize different ligands, but pattern recognition by TLRs triggers cytokine production by receptor-bearing cells [132]. When TNF-α receptors, IL-1 receptors, or Toll-like receptors are activated, they recruit IκB kinase (IKK), a complex composed of two serine/threonine protein kinases (IKKα and IKKβ) and a regulatory subunit called IKKγ. IKKγ is more commonly referred to as NF-κB essential modulator (NEMO). IKK phosphorylates IκB (one of the three isoforms, IκBα, IκBβ, or IκBε), inhibitor of NF-κB, and marks it for ubiquitination and subsequent proteasomal degradation. Subsequently, NF-κB translocates to the nucleus to turn on various genes that are involved in inflammation or the innate immune response. One of the genes turned on by NF-κB encodes IκBα, thus effectively regulating the signaling cascade through a negative feedback loop [16].

One TLR, TLR4, is unique as it recognizes a variety of nonbacterial agonists such as taxol [133], fibronectin [134, 135], and heat shock protein 60 [136] in addition to its known agonist lipopolysaccharide which is found in the cell walls of gram negative bacteria. UV exposure produced hyaluronic acid fragments (HAF) that promoted TLR4 translocation into the lipid rafts to initiate signaling. This trafficking is mediated, at least in part, by NAPDH oxidase-dependent ROS generation, which leads to the recruitment of a NF-κB subunit, p65, for encoding inflammatory molecules. This inflammatory machinery can be blocked by superoxide dismutase 3 [137].

TLR4 is also required for UV-induced immune suppression [138–140]. UVB-induced DNA damage is one of the earliest molecular events in UVB-induced immune suppression. CPD are repaired more efficiently in the skin and dendritic cells of TLR4-deficient mice. Furthermore, expressions of XPA and DNA repair cytokines (IL-12 and IL-23) are significantly higher in the skin and dendritic cells of TLR4-deficient mice in comparison to TLR4-proficient mice [141]. In animal models, TLR4 contributes to UV-induced cutaneous tumor development. UV-induced cutaneous carcinogenesis is retarded in TLR4-deficient mice.

8.14   AP-1 and the MAPK Signaling Pathway

AP-1 is another dimeric transcription factor activated by UV exposure. Members of Jun, Fos, musculoaponeurotic fibrosarcoma (MAF), and activating transcription factor (ATF) protein families form homodimers or heterodimers. Fos proteins cannot homodimerize, but they instead form stable heterodimers with Jun proteins. The main constituents of AP-1 in mammals belong to members of Fos (c-Fos, FosB, Fra-1, and Fra-2) and Jun (c-Jun, JunB, and JunD). Like NF-κB, the dimer composition of AP-1 dictates the types of genes that are regulated. Of the Jun proteins, c-Jun is the most potent activator of transcription. AP-1 is induced by various extracellular signals, which, in addition to UV irradiation, includes proinflammatory cytokines, genotoxic stress, and growth factors.

AP-1 induction depends on the MAPK signaling pathway. The MAPK signaling pathway, comprising several different protein kinases, helps relay the extracellular signal downstream to the nucleus via AP-1. Three family members of MAPKs are p38-MAPKs, extracellular-signal-regulated kinases (ERKs), and c-Jun amino-terminal kinases (JNKs). Subfamily members

exist for each of the three families: p38-MAPKs (isoforms α, β, γ, and δ), JNKs (JNK1, JNK2, and JNK3), and ERKs (ERK1/2). ERKs tend to respond to phorbol esters, many of which are tumor promoters, and growth factors (e.g., epidermal growth factor (EGF) binding to EGFR causes EGFR internalization and subsequently triggers ERK1/2 activity) [142, 143]. On the other hand, p38-MAPKs and JNKs tend to respond more to stress stimuli, such as proinflammatory cytokines and UV [144, 145]. Thus, AP-1 induction during photocarcinogenesis is largely mediated through JNKs and p38-MAPKs [146]. UVB irradiation clusters or causes multimerization of cell surface receptors that trigger the MAPK signaling pathway via JNK activation [147]. Both UVA and UVB have the ability to induce the p38 MAPK cascade, and this activation occurs through rapid phosphorylation of p38 MAPK, rather than through increased p38 protein synthesis [148, 149]. The effect of UVB on *c-fos* and *c-jun*, which are downstream targets in the MAPK signaling pathway and directly code for AP-1 constituents, has also been demonstrated by increased mRNA expression of both genes following administration of sublethal doses of UVB to human keratinocytes. Pretreatment with an antioxidant, N-acetylcysteine, inhibits UVB-induced JNK activation. Thus, ROS formation plays a crucial role in signal transduction activation during the UV response [150].

8.15   PI3K Signaling Pathway and COX-2 Activation

COX catalyzes the rate-limiting first step in generating prostaglandins, prostacyclins, and thromboxanes, all of which are derived from arachidonic acid. Two isoforms of COX are recognized: COX-1 and COX-2. COX-1, the dominant isoform, is expressed constitutively in most cells and is generally considered a housekeeping enzyme involved in such functions as cytoprotection of the stomach lining. COX-2 is undetectable in healthy adult skin tissue, but is inducible by UV and is upregulated by cytokines and growth factors. PGE_2, the primary product of COX-2 in the skin, induces proliferation, prevents apoptosis, promotes inflammation, and causes angiogenesis. Increased COX-2 has been detected in AKs, SCCs, and BCCs [151–153]. At least two separate UV-induced COX-2 activation pathways exist. These pathways include the ROS-mediated ligand-independent activation of EGFR, the p53-driven induction of heparin-binding EGF (HB-EGF), and the activation of AhR. All three pathways involve EGFR activation and its downstream signaling cascades. EGFR can cause activation in the phosphoinositide 3-kinase (PI3K) signaling pathway, in addition to the MAPK signaling pathway activation. Mammalian PI3Ks are lipid kinases that evolved from a single enzyme conserved in all eukaryotes. Three true PI3Ks, classes I–III, have been identified. Class IV PI3K-related kinases (mTOR, ATM, ATR, and DNA-PK), which are serine/threonine kinases that phosphorylate proteins instead of lipids, have also been characterized [154]. UVB rapidly triggers p38 MAPK and PI3K signaling pathways, as shown in both SKH-1 mice and HaCaT cells [155–157]. Thus, p38 MAPK is implicated in both AP-1 induction and COX-2 activation. In UVB-irradiated SKH-1 mice, SB202190 [4-(4-fluorophenyl)-2-(4-hydroxyphenyl)-5-(4-pyridyl)1H–imidazole], a topical inhibitor of p38 MAPK, decreases AP-1 activation, as measured by AP-1 luciferase reporter activity, by 84%; LY294002 [2-(4-morpholinyl)-8-phenyl-4H-1-benzopyran-4-one], a topical inhibitor of PI3K, decreases AP-1 activation by 68%. Both inhibitors decrease COX-2 expression as measured by western blot analysis [155]. The importance of ROS generation, an ability attributed more to UVA than UVB, in COX-2 induction is highlighted by the role that UVA can play in COX-2 expression. An estimated 70% of COX-2 induction by sunlight is due to UVA [158, 159].

8.16 Ornithine Decarboxylase and the Polyamine Biosynthetic Pathway

Polyamines have been implicated in skin carcinogenesis since it was observed that tumor promoting agents augment ornithine decarboxylase (ODC) activity. ODC is the rate-limiting enzyme in the polyamine biosynthetic pathway. The conversion of L-ornithine to putrescine ultimately results in an elevated production of spermine [160]. Spermine causes cell growth and proliferation of mutant keratinocytes. Chronic UVB exposure for up to 27 weeks in SKH:HR-1 elevated basal levels of ODC in epidermis up to 350-fold compared to age-matched control mice [161]. In addition, oral administration of α-difluoromethylornithine (DFMO), a suicidal ODC inhibitor, to UVB-irradiated PTCH1+/− mice reduced the number of UVB-induced NMSCs to about 40% of that observed in the vehicle-treated control group [162]. In human studies, oral administration of DFMO to patients with actinic damage for up to 4 years resulted in greater than 30% reduction in BCCs [163].

8.17 Photoimmunology

The genesis of photoimmunology, interactions between photons and the immune system, stemmed from experiments that sought to understand the mechanisms behind UVB-induced skin cancer [164, 165]. In 1974, Kripke reported that skin cancers that develop in UV-irradiated mice cannot be transplanted to immunocompetent age- and sex-matched syngeneic recipient mice. Instead, transplanted UV-induced tumors only grow in immunocompromised mice, and adding lymphocytes to the immunocompromised hosts mitigates tumor growth [166]. A subsequent study revealed that mice subjected to subcarcinogenic doses of UVR also failed to reject highly antigenic, transplanted UV-induced tumors [167]. Thus, it was suggested that suppression of the immune system is required for photocarcinogenesis since skin tumors that develop after UV exposure are highly antigenic and need to grow over an extended period of time [168]. A lack of immunologic destruction of these highly antigenic tumors suggested that UVR causes a breakdown in immunosurveillance. Furthermore, the immunosuppressive effect exerted by UVB is systemic since intravenous injection of UV-induced fibrosarcomas yields more pulmonary metastases in UV-irradiated normal synge-neic mice in a dose-dependent manner than those in untreated control mice [169].

8.17.1 UVR and Cell-Mediated Immune Reactions

The mechanistic data supporting the immunosup-pressive properties of UVR were largely gener-ated by studying the effect of UV on the induction of delayed-type hypersensitivity (DTH) and con-tact hypersensitivity (CHS). DTH and CHS, both T-cell-mediated hypersensitivity reactions, pro-vide a method of generating quantifiable results that reflect the cell-mediated immune function in mice and other rodent species. CHS and DTH reactions are produced by antigen-specific effec-tor T-cells and can be adoptively transferred by purified T-cells to recipient mice from previously sensitized, genetically identical donor mice.

UVB impairs induction of both CHS and DTH. Mice sensitized with the experimental hap-tens, dinitrofluorobenzene (DNFB) or trinitro-chlorobenzene (TNCB), 3–15 days following a single dose of UVB radiation show a dose-dependent suppression of CHS [164]. Impairment in DTH and CHS following UVB irradiation has been demonstrated in multiple studies using mice and guinea pigs [170–176].

The mechanistic basis behind the immunosup-pressive properties of UVB can be explained by three categories of UVB-driven effects: direct impairment of the antigen-presenting cell (APC) function, induction of antigen-specific T regula-tory cells (Treg cells, formerly known as suppres-sor T-cells), and changes in the production of soluble mediators that regulate APCs and Treg cells.

8.17.2 Direct Impairment of APC Function

Cutaneous exposure to UVR impairs APC func-tion by destroying the dendritic network in the skin. During cutaneous exposure to haptens or antigens, cutaneous dendritic cells "take up" and process antigens during their migration to regional lymph nodes. Originally, Langerhans cells were considered the primary antigen-presenting cells for effector T-cells. More recent studies have provided evidence that this may not be the case [177–179]. Rather, dendritic cells located in the dermis may be the primary antigen-presenting cells for cutaneous CHS and DTH reactions. Since antigen presentation is an obliga-tory step for the initiation of cell-mediated immune responses, the discovery that UVR impairs the antigen-presenting function of den-dritic cells generated a considerable amount of interest. Toews and colleagues [176] observed that induction of CHS by DNFB was impaired when attempting to sensitize mice through skin that had previously been exposed to UVB radia-tion. The immunosuppression observed was both long-term, implying an induction of immunotol-erance, and antigen-specific, and mediated by Treg cells [170].

8.17.3 Induction of Antigen-Specific T Regulatory Cells

UVR causes active induction of antigen-specific immunosuppression, and it does so through the unperturbed activation of Treg cells that suppress immune responses. As mentioned previously, UV-induced tumors are highly antigenic and their transplantation is only possible in immunosup-pressed or UVB-irradiated mice. When T-cells from mice subjected to UVR were transferred to syngeneic recipients, UV-induced tumors grew progressively, indicating that Treg cells have been generated that facilitate the growth of UV-induced tumors [180]. Interestingly, UV irradiation does not affect the ability to reject non-UV-induced tumors in mice. Thus, UV irradiation was said to confer selective, not general, unresponsiveness against UV-induced tumors [181, 182].

Characterization of UVR-Tregs reveals two dis-tinct lineages of cells. One expresses phenotypic markers CD4, CD25, transcription factor Foxp3 and a negative regulatory marker, CTLA-4 (CD152). Other identified surface markers include GITR, neuropilin, and CD62L [183]. A second group of antigen-specific immunosuppressive cells expresses CD3, a T-cell marker, and DX5 (CD49b), a natural killer (NK) cell marker. Accordingly, they are called NKT cells [184]. Recently, increased expression of RANKL on UVB-irradiated mouse keratinocytes has been reported along with an increase in $CD4^+CD25^+$ T-cells in draining lymph nodes [185]. Thus, keratinocytes may play a role in induction of UVR-Tregs aside from their characterized role of cytokine release.

8.17.4 Production of Soluble Immune Mediators

UVR-induced Tregs and UV-irradiated keratinocytes facilitate immunosuppression by releasing cytokines and other soluble mediators capable of suppressing cell-mediated immune responses.

$CD4^+CD25^+CTLA4^+FoxP3^+$ UVR-Tregs exert their effect by secreting IL-10, which is immunosuppressive. $CD11b^+$ macrophages, which populate the skin following UVB exposure, have been identified as other sources of IL-10 [186–189]. The NKT subset's immunosuppressive action depends on its release of IL-4, a T_H2 cytokine that suppresses both DTH and antitumor activity following UV exposure [190]. TLR3, a cell surface receptor found on keratinocytes, has also been implicated in immunosuppression mediated by UVB [138, 164]. TNF-α appears to exert its immunosuppressive effect through its interaction with the membrane receptor TNFR2 [191].

TLRs play an important role in the UV response. The signals that are transmitted through TLRs result in increased amounts of IL-10 and TNF-α. TLR3, a cell surface receptor found on keratinocytes, has also been implicated in immunosuppression mediated by TNF-α [192].

8.17.5 DNA Is the Chromophore for UVB-Induced Immunosuppression

Current evidence supports the concept that DNA is the chromophore for UV-induced immunosuppression. Based on variety of observations:

(i) the action spectra for DNA damage and UV-induced immune suppression are remarkably similar [193]; (ii) in animal models, exogenous administration of DNA repair enzymes reverses the immunosuppressive effects of UVB [194]; (iii) the cytokines IL-12, IL-23, and IL-18, all of which augment the induction of enzymes involved in NER and enhance the removal of CPDs from UV-irradiated skin, prevent UV-mediated suppression of CHS in mice [195–200]; (iv) patients with XP, a disease in which there is an inherited defect in DNA repair, also have impaired DTH responses, reduced circulating CD4/CD8 T-cell ratios, defective NK cell function, and impaired production of interferon-γ [201]; (v) UVR stimulates epidermal production of *cis*-urocanic acid, platelet activating factor, and serotonin, all of which have been shown to have immunosuppressive effects. These agents produce these effects by interfering with repair of DNA damage [168, 202].

8.17.6 Immunosuppression: A Risk Factor for Photocarcinogenesis in Humans

Suppression of the immune system plays a major role in photocarcinogenesis in humans. For example, HIV patients have an increased risk of NMSC (adjusted rate ratio 2.1, 95% CI = 1.9–2.3) that increases inversely with lower CD4 counts [203]. Organ transplant patients who are chronically immunosuppressed exhibit a 65- to 250-fold increased likelihood of SCC development and 10-fold increased likelihood of BCC development [204]. The longer the course of immunosuppressive therapy, the greater the risk of NMSC development. Moreover, SCCs in immunosuppressed patients have a higher rate of metastasis compared to the general population [205, 206]. Furthermore, azathioprine and cyclosporine have both been shown to inhibit the rate of CPD repair [207, 208].

Similarly, patients with lymphoma, who also have suppressed immune responses, are at increased risk for NMSCs [209]. Subtle immunological defects have been identified in otherwise healthy individuals who have had skin cancers. They have diminished reactions to skin test antigens and reduced sensitization to the experimental contact sensitizing agent dinitrochlorobenzene [140, 210]. Increased Treg cells are present around tumor islands in BCCs [211]. Also, individuals treated with PUVA have reduced immunization rates to contact allergens [212] and fewer circulating peripheral CD4+ T-cells [213, 214].

8.18 Therapeutic Options

A primary goal of research regarding photocarcinogenesis is to develop better methods for treatment and prevention of these neoplasms.

8.18.1 Currently Available Treatment Options for NMSC and Its Precursor Lesions

Treatment options that are currently available for NMSC are primarily surgical: surgical excision, Mohs micrographic surgery or electrodessication and curettage. However, in patients who are poor candidates for surgery or for lesions in which operative outcomes would be inferior, a number of nonsurgical options are available. These include

topical application of imiquimod, cryotherapy, and radiotherapy. Recently, two medical treatments for advanced and metastatic BCC, vismodegib and sonigeb, have received FDA approval. Both of these agents are small molecule inhibitors of the hedgehog pathway. It is important to note that most nonsurgical forms of therapy are less effective than surgical removal. A detailed discussion of current treatment modalities for NMSCs is beyond the scope of this review. The reader is referred to the other references for a more detailed discussion of current treatment modalities for AKs and NMSCs [215–220] (Table 8.1).

8.18.2 Chemoprevention of Photocarcinogenesis

Because of the incidence, cost, and morbidity associated with the treatment of NMSC, there has been an attempt to identify methods for their prevention. Since UV exposure is the largest modifiable risk factor for photocarcinogenesis, there has been a concerted effort by various healthcare organizations to educate the public about the hazards of overexposure to the sun and artificial UV light sources. This includes counseling patients to avoid excessive sun and to discontinue the use of tanning beds, to avoid outdoor activities during peak hours of UV intensity

Table 8.1 Selected investigative treatments of NMSCs and precursor lesions

Name	Proposed indication	Mechanism	Status	References
Resiquimod[a]	AK	TLR7 and TLR8 agonist. A more potent inducer of IL-12, TNF-α and INF-γ in vitro than imiquimod	Phase II (completed)	[221]
Dz13[b]	BCC	Deoxyribozyme (DNAzyme) that binds and cleaves *c-jun* mRNA	Phase I (completed)	[222, 223]
Itraconazole	BCC	Hedgehog signaling pathway inhibitor. Thought to act as an inverse agonist of SMO and prevent its ciliary accumulation	Phase II (completed)	[224, 225]
Cetuximab	SCC	Anti-EGFR monoclonal antibody	Phase II (completed)	[226–228]
Panitumumab	SCC	Anti-EGFR monoclonal antibody	Phase II (completed)	[229, 230]
Gefitinib	SCC	EGFR tyrosine kinase inhibitor	Phase II (completed)	[231]

[a]Investigated as a topical agent
[b]Investigated as an intratumorally injected agent
Investigated *as a systemic agent unless noted otherwise*

(i.e., 10 a.m.–4 p.m.), to conduct outdoor activities in shaded areas if possible, to wear protective clothing, and to apply sunscreens. Because the incidence of NMSC continues to rise despite these methods for prevention, identification of chemopreventive agents is an active area of research. Chemoprevention refers to prevention of cancer through dietary or pharmacological intervention. Because these agents must be administered on a chronic basis to individuals at risk, the ideal chemopreventive agent would be one that is highly effective, has little or no toxicity, is easy to administer, has low cost, and is acceptable to individuals at risk [232].

8.18.2.1 Sunscreens and Other Photoprotective Measures

Sunscreens, available since 1928, are safe and widely available. The sunscreen industry initially focused on increasing the sun protection factor (SPF) and UVB protection for its products. More recently, the emphasis has shifted to increasing UVA protection since UVA exposure also plays a significant role in photocarcinogenesis. UVB protection is rated by SPF. Countries have different ways of grading UVA protection (e.g., photoprotection factor of UVA (PFA) in Japan; Boots star rating in the UK and Ireland; critical wavelength method in the USA). Organic ingredients and inorganic ingredients (e.g., titanium dioxide and zinc oxide) are mixed and placed in a vehicle to produce a sunscreen product that offers broad-spectrum protection [233].

Unfortunately, sunscreens provide only moderate protection against SCCs, and there is almost no evidence that they reduce the incidence of BCCs [234–238]. In addition, a cross-sectional study of 1034 subjects, with a follow-up after 1–2 years that evaluated the presence and changes in AKs/SCCs, found no protective effect from sunscreen use or protective clothing when controlling for confounders [239]. In addition, purchase of sunscreen per head in the USA and UK has increased since the 1990s while the incidence of NMSCs has continued to rise, leading many to question the effectiveness of sunscreen in NMSC prevention [240].

8.18.2.2 Investigative Chemopreventive Agents Against Photocarcinogenesis

The need for alternatives to sunscreens and other photoprotective measures has been acknowledged by healthcare researchers from many different countries for a number of years. Both drugs approved for other indications and dietary supplements have been investigated and are listed in Tables 8.2 and 8.3. There have been several agents that have been evaluated in humans for their chemopreventive activity against skin cancer.

Table 8.2 Selected investigative chemopreventive agents against photocarcinogenesis (drugs)

Name	Mechanism	Status	References
Celecoxib	Reversible selective COX-2 inhibitor. Nonsteroidal anti-inflammatory drug (NSAID)	Phase III(completed)	[241, 242]
Tretinoin[a]	Vitamin A derivative. Anti-proliferative, and capable of normalizing follicular epithelial differentiation and keratinization	Phase III, terminated due to an increase in all-cause mortality in treatment group	[243]
Acitretin	Vitamin A derivative and a monoaromatic retinoid. Anti-proliferative, and capable of normalizing follicular epithelial differentiation and keratinization	Phase III (completed) (ClinicalTrials.gov Identifier: NCT00644384 and NCT00003611)	[244]
T4N5[a]	Pyrimidine dimer-specific DNA glycosylase. Binds CPD and carries out base excision at its 5′ end (base excision repair).	Phase II to prevent NMSC recurrence in renal transplant patients (ClinicalTrials.gov Identifier:NCT00089180)	[245–249]
DFMO	ODC inhibitor. Decreases concentration of polyamines in tissues	Phase III (completed)	[163]

[a]Investigated as a topical agent
Investigated *as a systemic agent unless noted otherwise*

Table 8.3 Selected investigative chemopreventive agents against photocarcinogenesis (dietary supplements)

Name	Mechanism	Status	References
Grape constituents (e.g., proantho-cyanidins, resveratrol)	Reduction of UVB-induced lipid peroxidation, increased apoptosis, increased CPD repair, enhanced cell-mediated immune response in a TLR-4 dependent manner, inhibition of CD4$^+$CD25$^+$Tregs function, inhibition of TGF-β synthesis, increased levels of IL-2, IL-12 and IFN-γ	Preclinical	[250–256]
Green tea polyphenols (e.g., (−)-epigallocatechin-3-gallate)	Induces NER and inhibits UV-induced immunosuppression in an IL-12 dependent manner. Inhibits angiogenic factors: MMPs and VEGFs. Inhibits induction of AP-1, ODC and COX-2 following UVB exposure	Preclinical	[251, 257]
Pomegranate	Fruit from *Punica granatum*. Contains hydrolyzable tannins, polyphenols, and anthocyanidins that act as free-radical scavengers. Reduces CPD and 8-oxoG formation. Inhibits MMP-2, MMP-9, COX-2, and ODC. Inhibits UVB-mediated phosphorylation of JNK1/2, ERK1/2, and p38. Also inhibits UVB-mediated phosphorylation of MAPK and activation of NF-κB	Preclinical	[258–264]
Nicotinamide (vitamin B3)	Prevent UV-induced ATP depletion to fuel cellular processes such as DNA repair that requires a high amount of ATP. Substrate for poly-adenosine diphosphate ribose polymerases (PARPs), molecular sensors of double and single DNA strand breaks. PARPs are required to maintain genomic stability	Phase III (Australian New Zealand Clinical Trials Registry: ACTRN12612000625875)	[265–273]
Silymarin	Milk thistle derivative. Inhibits UV-induced immunosuppression through NER induction. Inhibits UVB-induced oxidative stress via inhibition of infiltrating CD11b + cells (activated macrophages and neutrophils). Upregulates secretion of IL-2 and IFN-γ. Downregulates UVB-induced NF-κB signaling pathway, COX-2 expression, IL-10 secretion	Preclinical	[274–276]
Polypodium leucatomos	Tropical fern leaf derivative. Reduces CPD formation. Downregulates UV-induced production of nitric oxide and TNF-α, as well as activation of AP-1 and NF-κB. Inhibit MMP-1, 2, 3, and 9, COX-2. Enhances p53 expression following UV irradiation	Preclinical	[277–282]

Celecoxib

Celecoxib, a selective COX-2 inhibitor, has been on the US market since 1998 to treat inflammatory conditions such as osteoarthritis, rheumatoid arthritis, ankylosing spondylitis, and juvenile idiopathic arthritis. A randomized, double-blind, placebo-controlled study of celecoxib has been completed. Two hundred forty subjects with

10–40 AKs and a prior histological diagnosis of at least one AK or NMSC were randomly assigned to take celecoxib or placebo for 9 months and were followed up for an additional 2 months. At 11 months, those in the celecoxib arm of the trial show fewer NMSCs than those in the placebo group (rate ratio .41, 95% CI = 0.23–0.72). The findings were statistically significant ($p = 0.002$). Results also showed a lower number of SCCs with a rate ratio of .42 (95% CI = 0.19–0.93) and BCCs with a rate ratio of .40 (95% CI = 0.18–0.93) that was also statistically significant ($p = 0.032$). The rate of serious adverse events and cardiovascular events was not significantly different between two groups [242]. Although the results are promising, the feasibility of this approach must be balanced with its potential consequences since long-term administration of COX-2 inhibitors has been linked to an increased risk of serious cardiovascular and gastrointestinal events [283–285].

α-Difluoromethylornithine (DFMO)

DFMO is an ODC inhibitor that decreases the concentration of polyamines in tissues and has been investigated as a preventative agent for NMSC in a randomized, double-blind, placebo-controlled study involving 291 subjects with a prior history of NMSCs. The primary endpoint of new NMSCs did not differ significantly ($p = 0.069$) between the DFMO group (260 new NMSCs) and the placebo group (363 new NMSCs). However, the difference in new BCCs was statistically significant ($p = 0.03$). Mild clinically inapparent ototoxicity was also noted in the DFMO group as a significant adverse effect [163].

Topical Tretinoin

The Veterans Affairs Topical Tretinoin Chemoprevention (VATTC) Trial was a randomized, blinded, placebo-controlled study involving 1131 subjects (mean age = 71) with at least two prior NMSCs evaluating the efficacy of topical 0.1% tretinoin as a possible chemopreventive agent. Retinoids are a class of natural or synthetic compounds with vitamin A activities and possess anti-proliferative and pro-apoptotic effects. Identified new BCCs and SCCs were the primary outcomes in the VATTC Trial. Unfortunately, there was no statistically significant difference between the tretinoin and placebo-treated groups for BCC and SCC. Moreover, the trial was stopped prematurely due to increased all-cause mortality in the intervention group (hazard ratio 1.54, 82 vs. 53 deaths) that was statistically significant ($p < 0.01$). No reasonable explanation was found for the increased mortality despite extensive analysis [243].

Oral Retinoids

Oral administration of retinoids has been shown to be remarkably effective at suppressing NMSC development in renal transplant recipients [286, 287] and in XP patients [288]. In otherwise healthy individuals at high risk for skin cancer, an early study using high dose isotretinoin (an average of 3.1 mg/kg/day) for 8 months to patients with existing skin cancers resulted in a marked decrease in new skin cancers in some patients while on treatment [289]. However, at these high doses, toxicity limited their use. Tangrea et al. examined a lower dose of isotretinoin (10 mg daily for 3 years) [290]. Compared to placebo, there was no reduction in the incidence of BCCs. Newer retinoids continue to be studied for skin cancer prevention. The acetylated retinoid tazarotene has been shown in animal models to profoundly reduce the onset of new basal cell carcinomas due to ultraviolet and ionizing radiation [291]. These findings were consistent with an open label trial in humans, which showed that 16 of 30 sporadic BCCs regressed after 8 months of treatment with topical tazarotene [292]. In other studies, acitretin at a dose of 25 mg five times per week over 2 years in individuals who had had at least two BCCs or SCCs did not result in a significant reduction in new NMSCs [244]. Because retinoids suppress tumor growth in multiple tissues, the search for vitamin A derivatives with lower toxicity continues [293–295].

T4 Endonuclease V

Another notable agent that has been investigated in humans is topical application of bacteriophage T4 endonuclease V (T4N5). T4N5 is a pyrimidine dimer-specific DNA glycosylase that binds CPD and carries out base excision at its 5′ end. Thus, topical T4N5 increases base excision repair induction of damaged DNA. A prospective, multicenter, double-blind study was carried out for a year in 30 XP patients. Twenty were assigned T4N5 liposome lotion and 10 were assigned a placebo lotion. The treatment group had a 68% reduction in mean annual rate of new AKs and a 30% reduction in the mean annual rate of new BCCs. Reductions of BCCs and AKs in the treatment group were both statistically significant. No significant adverse effects were noted in the treatment group [248].

Nicotinamide

Nicotinamide is a precursor of nicotinamide adenine dinucleotide (NAD), a ubiquitous coenzyme required for ATP synthesis in the Kreb Cycle. Nicotinamide has been shown in in vitro studies to augment DNA repair processes which require large amounts of ATP [270], and prevent photocarcinogenesis and photoimmunosuppression in mice [265, 268, 269].

These findings were extended in the ONTRAC (Oral Nicotinamide to Reduce Actinic Cancer) trial, which is a phase III double-blind, randomized clinical trial investigating the efficacy of oral nicotinamide in preventing NMSCs. Nicotinamide is derived from niacin (nicotinic acid or vitamin B3). ONTRAC involved 386 immunocompetent subjects with ≥2 NMSCs. The subjects were randomized to receive oral nicotinamide 500 mg twice a day or placebo for 12 months. The relative rate reduction (RRR) between treatment groups was 0.23 (95% CI 0.04–0.38) and was statistically significant ($p = 0.02$) when adjusting for NMSC history and study site. Reductions in both BCCs (RRR = 0.20, 95% CI = −0.06–0.39) and SCCs (RRR = 0.30, 95% CI = 0–0.51) were observed. However, only the reduction in SCCs was statistically significant ($p = 0.05$). The number of AKs was lower in the treatment group by 13% at 12 months ($p < 0.005$). No difference in adverse events was noted [296].

8.19   Summary (Fig. 8.7)

Photocarcinogenesis, driven by sunlight, underlies the pathogenesis of both BCCs and SCCs. Three stages of photocarcinogenesis, initiation,

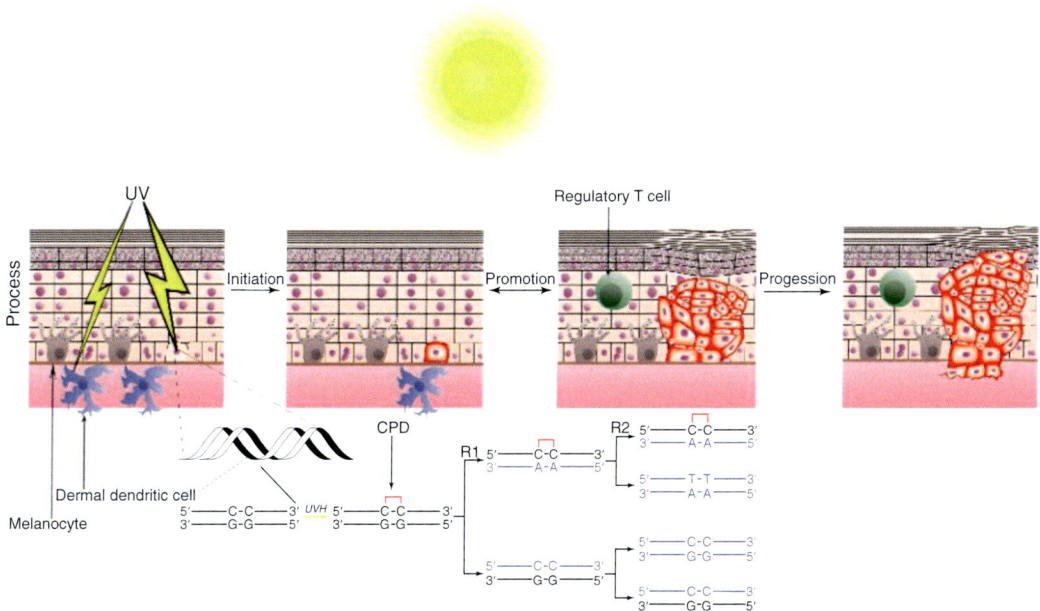

Fig. 8.7 Summary of photocarcinogenesis. *UV* ultraviolet, *CPD* cyclobutane pyrimidine dimer, *R1* round 1 of DNA replication, *R2* round 2 of DNA replication

promotion, and progression, have been described. During initiation, UV, largely UVB, induces DNA damage in keratinocytes, primarily in the form of CPDs and 6-4PPs. When unrepaired, DNA containing these lesions can be replicated and the resulting copy will contain single or tandem base substitutions called UV signature mutations (C → T or CC → TT). These lesions are significant since they may cause mutations in genes that control cellular proliferation, regulation, or differentiation. *TP53* (in SCC and BCC) and *PTCH* (in BCC) constitute two loci that contain key initiating mutations of photocarcinogenesis. During promotion, repeated doses of UV cause chronic inflammation that encourages clonal expansion of initiated cells. UV-induced activation of transcription factors, NF-κB and AP-1, trigger signaling pathways that increase inflammation and promote proliferation of the mutant keratinocytes. The result of promotion is an AK, a reversible premalignant lesion. During progression, premalignant cells gain genetic instability and changes associated with cancerous cells (i.e., acquired chromosomal aberrations). Premalignant cells increase their migratory capacity by undergoing EMT. The induction of angiogenesis is also essential during this stage. Induction of ODC and COX-2 are also implicated in both promotion and progression. The end product of progression is an invasive SCC. BCCs are believed to develop de novo without a precursor lesion.

Progressing through all three stages of photocarcinogenesis usually takes years to decades because initiation is rapid, but promotion and progression are slow and rate-limiting. Only a small portion of UV-induced photoproducts cause mutations since repair and bypass mechanisms both exist. The importance of excision repair is highlighted in patients with XP, which is caused by a germline mutation in any one of its seven different complementation groups of NER. XP patients develop actinically damaged skin and NMSCs in sun-exposed areas as early as 3–5 years of age and possess a significantly reduced life span, as well as many other comorbidities.

UV-induced defects in immunological surveillance contribute to the development of skin cancer. Repeated UV exposures inactivate immunological defenses and allow neoplastic cells to grow and develop into invasive skin cancers. UV-irradiated keratinocytes produce soluble immunosuppressive mediators. UV also causes direct impairment of APC function by dermal dendritic cells, and induction of antigen-specific Treg cells.

Currently available treatment options for NMSC are primarily surgical. Nonsurgical options are also available, but they are less effective. These include topical application of imiquimod, cryotherapy, and radiotherapy. Systemic treatments for advanced and metastatic BCC, vismodegib and sonigeb, have recently received FDA approval.

Although widely available, sunscreens provide only moderate protection against AKs and SCCs. Thus, there is a need for alternatives to sunscreens and other photoprotective measures. Investigative chemopreventive agents against photocarcinogenesis include: celecoxib, DFMO, T4N5, and oral nicotinamide. Dietary supplements, such as grape constituents, green tea polyphenols, and pomegranate, are also being investigated. Despite the rising incidence of NMSC, many promising chemopreventive targets against photocarcinogenesis exist due to recent advances and they warrant further investigation in the future.

References

1. Rogers HW, Weinstock MA, Feldman SR, Coldiron BM. Incidence estimate of nonmelanoma skin cancer (keratinocyte carcinomas) in the US population, 2012. JAMA Dermatol. 2015;151(10):1081–6. doi:10.1001/jamadermatol.2015.1187.
2. American Cancer Society. Cancer facts & figures 2015. Atlanta: American Cancer Society; 2015.
3. Leiter U, Garbe C. Epidemiology of melanoma and nonmelanoma skin cancer—the role of sunlight. In: Reichrath J, editor. Sunlight, vitamin D and skin cancer. New York, NY: Springer; 2008. p. 89–103.
4. Christenson LJ. Incidence of basal cell and squamous cell carcinomas in a population younger than 40 years. JAMA. 2005;294:681. doi:10.1001/jama.294.6.681.

5. Collins GL, Nickoonahand N, Morgan MB. Changing demographics and pathology of non-melanoma skin cancer in the last 30 years. Semin Cutan Med Surg. 2004;23:80–3.
6. Niederhuber JE, Armitage JO, Doroshow JH, et al. Abeloff's clinical oncology. 5th ed. Philadelphia, PA: Elsevier; 2014.
7. Rhee JS, Matthews BA, Neuburg M, et al. Creation of a quality of life instrument for nonmelanoma skin cancer patients. Laryngoscope. 2005;115:1178–85. doi:10.1097/01.MLG.0000166177.98414.5E.
8. Guy GP, Machlin SR, Ekwueme DU, Yabroff KR. Prevalence and costs of skin cancer treatment in the U.S., 2002−2006 and 2007−2011. Am J Prev Med. 2015;48:183–7. doi:10.1016/j.amepre.2014.08.036.
9. Altmeyer P, Hoffmann K, Stücker M, editors. Skin cancer and UV radiation. Berlin, NY: Springer; 1997.
10. Norval M, Kellett P, Wright CY. The incidence and body site of skin cancers in the population groups of South Africa: skin cancers in South Africa. Photodermatol Photoimmunol Photomed. 2014;30:262–5. doi:10.1111/phpp.12106.
11. Madan V, Lear JT, Szeimies R-M. Non-melanoma skin cancer. Lancet. 2010;375:673–85. doi:10.1016/S0140-6736(09)61196-X.
12. Goldsmith LA, Fitzpatrick TB. Fitzpatrick's dermatology in general medicine. New York: McGraw-Hill Medical; 2012.
13. Urbach F. The historical aspects of photocarcinogenesis. Front Biosci. 2002;7:e85–90.
14. Findlay GM. Ultra-violet light and skin cancer. CA Cancer J Clin. 1979;29:169–71.
15. Roffo A. Carcinomes et Sarcomes provoques par l'action du Soleil in toto. Bull Cancer. 1934;23:590–616.
16. Alberts B, editor. Molecular biology of the cell. 5th ed. Garland Science: New York; 2008.
17. Knudson AG. Hereditary cancer: two hits revisited. J Cancer Res Clin Oncol. 1996;122:135–40. doi:10.1007/BF01366952.
18. Schwab M (ed) (2011) Encyclopedia of cancer , 3rd ed. Springer, Heidelberg, NY.
19. National Institute of Environmental Health Sciences. National toxicology program (U.S.). 2014. Report on carcinogens.
20. Lim HW, Hönigsmann H, Hawk JLM. Photodermatology. New York: Informa Healthcare USA; 2007.
21. Strickland PT. Photocarcinogenesis by near-ultraviolet (UVA) radiation in Sencar mice. J Invest Dermatol. 1986;87:272–5.
22. Willis I, Menter JM, Whyte HJ. The rapid induction of cancers in the hairless mouse utilizing the principle of Photoaugmentation. J Invest Dermatol. 1981;76:404–8. doi:10.1111/1523-1747.ep12520945.
23. Cleaver JE. Defective repair replication of DNA in xeroderma pigmentosum. Nature. 1968;218(5142):652–6.
24. Cleaver JE. Xeroderma pigmentosum: a human disease in which an initial stage of DNA repair is defective. 1969;63(2):428–35.
25. Shacter E, Weitzman SA. Chronic inflammation and cancer. Oncology (Williston Park). 2002;16:217–226, 229. discussion 230–232.
26. Johnson TM, Rowe DE, Nelson BR, Swanson NA. Squamous cell carcinoma of the skin (excluding lip and oral mucosa). J Am Acad Dermatol. 1992;26:467–84.
27. Lee JM. The epithelial-mesenchymal transition: new insights in signaling, development, and disease. J Cell Biol. 2006;172:973–81. doi:10.1083/jcb.200601018.
28. Turro NJ, Ramamurthy V, Scaiano JC. Principles of molecular photochemistry: an introduction. Sausalito, CA: University Science Books; 2009.
29. Hsu TC, Young MR, Cmarik J, Colburn NH. Activator protein 1 (AP-1)- and nuclear factor kappaB (NF-kappaB)-dependent transcriptional events in carcinogenesis. Free Radic Biol Med. 2000;28:1338–48.
30. Pfafflin JR, Ziegler EN. Encyclopedia of environmental science and engineering. New York: Taylor & Francis; 2006.
31. Preedy VR. Aging oxidative stress and dietary antioxidants. Burlington: Elsevier Science; 2014.
32. Blum H. Carcinogenesis by ultraviolet light. Princeton: Princeton University Press; 1959.
33. Freeman RG. Data on the action spectrum for ultraviolet carcinogenesis. J Natl Cancer Inst. 1975;55:1119–22.
34. de Gruijl FR. Action spectrum for photocarcinogenesis. Recent Results Cancer Res. 1995;139:21–30.
35. de Gruijl FR, Sterenborg HJ, Forbes PD, et al. Wavelength dependence of skin cancer induction by ultraviolet irradiation of albino hairless mice. Cancer Res. 1993;53:53–60.
36. Black HS, deGruijl FR, Forbes PD, et al. Photocarcinogenesis: an overview. J Photochem Photobiol B. 1997;40:29–47.
37. de Laat A, van der Leun JC, de Gruijl FR. Carcinogenesis induced by UVA (365-nm) radiation: the dose-time dependence of tumor formation in hairless mice. Carcinogenesis. 1997;18:1013–20.
38. van Weelden H, de Gruijl FR, van der Putte SC, et al. The carcinogenic risks of modern tanning equipment: is UV-A safer than UV-B? Arch Dermatol Res. 1988;280:300–7.
39. Setlow RB, Grist E, Thompson K, Woodhead AD. Wavelengths effective in induction of malignant melanoma. Proc Natl Acad Sci U S A. 1993;90:6666–70.
40. Colantonio S, Bracken MB, Beecker J. The association of indoor tanning and melanoma in adults: systematic review and meta-analysis. J Am Acad Dermatol. 2014;70:847–857.e1–18. doi:10.1016/j.jaad.2013.11.050.
41. Wehner MR, Shive ML, Chren M-M, et al. Indoor tanning and non-melanoma skin cancer: systematic review and meta-analysis. BMJ. 2012;345:e5909.

42. de Gruijl FR, Van Der Meer JB, Van Der Leun JC. Dose-time dependency of tumor formation by chronic UV exposure. Photochem Photobiol. 1983;37:53–62. doi:10.1111/j.1751-1097.1983.tb04433.x.

43. Armstrong BK, Kricker A. The epidemiology of UV induced skin cancer. J Photochem Photobiol B. 2001;63:8–18.

44. Gallagher RP, Hill GB, Bajdik CD, et al. Sunlight exposure, pigmentary factors, and risk of nonmelanocytic skin cancer. I. Basal cell carcinoma. Arch Dermatol. 1995;131:157–63.

45. Diffey BL. Solar ultraviolet radiation effects on biological systems. Phys Med Biol. 1991;36:299–328.

46. Urbach F. Ultraviolet radiation and skin cancer of humans. J Photochem Photobiol B. 1997;40:3–7.

47. Lee SG, Ko NY, Son SW, et al. The impact of ozone depletion on skin cancer incidence in Korea. Br J Dermatol. 2013;169:1164–5. doi:10.1111/bjd.12472.

48. US Environmental Protection Agency. Updating ozone calculations and emissions profiles for use in the atmospheric and health effects framework model. Washington, DC: U.S. Environmental Protection Agency; 2015.

49. Hill DJ, Elwood JM, English DR, editors. Prevention of skin cancer. Dordrecht; Boston: Kluwer Academic Publishers; 2004.

50. Wehner MR, Chren M-M, Nameth D, et al. International prevalence of indoor tanning: a systematic review and meta-analysis. JAMA Dermatol. 2014;150:390. doi:10.1001/jamadermatol.2013.6896.

51. Cowan DO, Drisko RL. Elements of organic photochemistry. New York: Plenum Press; 1976.

52. Beukers R, Eker APM, Lohman PHM. 50 years thymine dimer. DNA Repair. 2008;7:530–43. doi:10.1016/j.dnarep.2007.11.010.

53. You YH, Lee DH, Yoon JH, et al. Cyclobutane pyrimidine dimers are responsible for the vast majority of mutations induced by UVB irradiation in mammalian cells. J Biol Chem. 2001;276:44688–94. doi:10.1074/jbc.M107696200.

54. Sinha RP, Häder DP. UV-induced DNA damage and repair: a review. Photochem Photobiol Sci. 2002;1:225–36.

55. Gould JW, Mercurio MG, Elmets CA. Cutaneous photosensitivity diseases induced by exogenous agents. J Am Acad Dermatol. 1995;33:551–73. quiz 574–576

56. Ley RD, Peak MJ, Lyon LL. Induction of pyrimidine dimers in epidermal DNA of hairless mice by UVB: an action spectrum. J Invest Dermatol. 1983;80:188–91.

57. de Gruijl FR, Van der Leun JC. Estimate of the wavelength dependency of ultraviolet carcinogenesis in humans and its relevance to the risk assessment of a stratospheric ozone depletion. Health Phys. 1994;67:319–25.

58. Young AR, Chadwick CA, Harrison GI, et al. The similarity of action spectra for thymine dimers in human epidermis and erythema suggests that DNA is the chromophore for erythema. J Invest Dermatol. 1998;111:982–8. doi:10.1046/j.1523-1747.1998.00436.x.

59. Bachelor MA, Bowden GT. UVA-mediated activation of signaling pathways involved in skin tumor promotion and progression. Semin Cancer Biol. 2004;14:131–8. doi:10.1016/j.semcancer.2003.09.017.

60. Mouret S, Baudouin C, Charveron M, et al. Cyclobutane pyrimidine dimers are predominant DNA lesions in whole human skin exposed to UVA radiation. Proc Natl Acad Sci. 2006;103:13765–70. doi:10.1073/pnas.0604213103.

61. Tewari A, Sarkany RP, Young AR. UVA1 induces cyclobutane pyrimidine dimers but not 6-4 photoproducts in human skin in vivo. J Invest Dermatol. 2012;132:394–400. doi:10.1038/jid.2011.283.

62. de Gruijl FR. Skin cancer and solar UV radiation. Eur J Cancer. 1999;35:2003–9.

63. Pettijohn DE, Hanawalt PC. Deoxyribonucleic acid replication in bacteria following ultraviolet irradiation. Biochim Biophys Acta. 1963;72:127–9. doi:10.1016/0926-6550(63)90324-4.

64. Seeberg E, Eide L, Bjørås M. The base excision repair pathway. Trends Biochem Sci. 1995;20:391–7.

65. Ruven HJ, Seelen CM, Lohman PH, et al. Strand-specific removal of cyclobutane pyrimidine dimers from the p53 gene in the epidermis of UVB-irradiated hairless mice. Oncogene. 1994;9:3427–32.

66. Bolognia J, Jorizzo JL, Schaffer JV. Dermatology. Philadelphia; London: Elsevier Saunders; 2012.

67. Burger A, Fix D, Liu H, et al. In vivo deamination of cytosine-containing cyclobutane pyrimidine dimers in E. coli: a feasible part of UV-mutagenesis. Mutat Res. 2003;522:145–56.

68. Choi J-H, Pfeifer GP. The role of DNA polymerase eta in UV mutational spectra. DNA Repair. 2005;4:211–20. doi:10.1016/j.dnarep.2004.09.006.

69. Lee D-H, Pfeifer GP. Deamination of 5-methylcytosines within cyclobutane pyrimidine dimers is an important component of UVB mutagenesis. J Biol Chem. 2003;278:10314–21. doi:10.1074/jbc.M212696200.

70. Song Q, Cannistraro VJ, Taylor J-S. Synergistic modulation of cyclobutane pyrimidine dimer photoproduct formation and deamination at a TmCG site over a full helical DNA turn in a nucleosome core particle. Nucleic Acids Res. 2014;42:13122–33. doi:10.1093/nar/gku1049.

71. Tu Y, Dammann R, Pfeifer GP. Sequence and time-dependent deamination of cytosine bases in UVB-induced cyclobutane pyrimidine dimers in vivo. J Mol Biol. 1998;284:297–311. doi:10.1006/jmbi.1998.2176.

72. Ziegler A, Leffell DJ, Kunala S, et al. Mutation hotspots due to sunlight in the p53 gene of

nonmelanoma skin cancers. Proc Natl Acad Sci U S A. 1993;90:4216–20.

73. Takasawa K. Chemical synthesis and translesion replication of a cis-syn cyclobutane thymine-uracil dimer. Nucleic Acids Res. 2004;32:1738–45. doi:10.1093/nar/gkh342.

74. Tessman I, Liu SK, Kennedy MA. Mechanism of SOS mutagenesis of UV-irradiated DNA: mostly error-free processing of deaminated cytosine. Proc Natl Acad Sci U S A. 1992;89:1159–63.

75. Stern RS. The risk of squamous cell and basal cell cancer associated with psoralen and ultraviolet a therapy: a 30-year prospective study. J Am Acad Dermatol. 2012;66:553–62. doi:10.1016/j.jaad.2011.04.004.

76. Karagas MR, Stukel TA, Umland V, et al. Reported use of photosensitizing medications and basal cell and squamous cell carcinoma of the skin: results of a population-based case-control study. J Invest Dermatol. 2007;127:2901–3. doi:10.1038/sj.jid.5700934.

77. Robinson SN, Zens MS, Perry AE, et al. Photosensitizing agents and the risk of nonmelanoma skin cancer: a population-based case-control study. J Invest Dermatol. 2013;133:1950–5. doi:10.1038/jid.2013.33.

78. Jensen AØ, Thomsen HF, Engebjerg MC, et al. Use of photosensitising diuretics and risk of skin cancer: a population-based case–control study. Br J Cancer. 2008;99:1522–8. doi:10.1038/sj.bjc.6604686.

79. Schmidt SAJ, Schmidt M, Mehnert F, et al. Use of antihypertensive drugs and risk of skin cancer. J Eur Acad Dermatol Venereol. 2015;29:1545–54. doi:10.1111/jdv.12921.

80. Kaae J, Boyd HA, Hansen AV, et al. Photosensitizing medication use and risk of skin cancer. Cancer Epidemiol Biomarkers Prev. 2010;19:2942–9. doi:10.1158/1055-9965.EPI-10-0652.

81. Cowen EW, Nguyen JC, Miller DD, et al. Chronic phototoxicity and aggressive squamous cell carcinoma of the skin in children and adults during treatment with voriconazole. J Am Acad Dermatol. 2010;62:31–7. doi:10.1016/j.jaad.2009.09.033.

82. Epaulard O, Saint-Raymond C, Villier C, et al. Multiple aggressive squamous cell carcinomas associated with prolonged voriconazole therapy in four immunocompromised patients. Clin Microbiol Infect. 2010;16:1362–4. doi:10.1111/j.1469-0691.2009.03124.x.

83. Ibrahim SF, Singer JP, Arron ST. Catastrophic squamous cell carcinoma in lung transplant patients treated with voriconazole. Dermatol Surg. 2010;36:1752–5. doi:10.1111/j.1524-4725.2010.01596.x.

84. McCarthy KL, Playford EG, Looke DFM, Whitby M. Severe photosensitivity causing multifocal squamous cell carcinomas secondary to prolonged Voriconazole therapy. Clin Infect Dis. 2007;44:e55–6. doi:10.1086/511685.

85. Vadnerkar A, Nguyen MH, Mitsani D, et al. Voriconazole exposure and geographic location are independent risk factors for squamous cell carcinoma of the skin among lung transplant recipients. J Heart Lung Transplant. 2010;29:1240–4. doi:10.1016/j.healun.2010.05.022.

86. Vanacker A, Fabré G, Van Dorpe J, et al. Aggressive cutaneous squamous cell carcinoma associated with prolonged voriconazole therapy in a renal transplant patient. Am J Transplant. 2008;8:877–80. doi:10.1111/j.1600-6143.2007.02140.x.

87. Miller DD, Cowen EW, Nguyen JC, et al. Melanoma associated with long-term voriconazole therapy: a new manifestation of chronic photosensitivity. Arch Dermatol. 2010a;146:300–4. doi:10.1001/archdermatol.2009.362.

88. Wikonkal NM, Brash DE. Ultraviolet radiation induced signature mutations in photocarcinogenesis. J Investig Dermatol Symp Proc. 1999;4:6–10.

89. Yin Y, Tainsky MA, Bischoff FZ, et al. Wild-type p53 restores cell cycle control and inhibits gene amplification in cells with mutant p53 alleles. Cell. 1992;70:937–48.

90. Ziegler A, Jonason AS, Leffell DJ, et al. Sunburn and p53 in the onset of skin cancer. Nature. 1994;372:773–6. doi:10.1038/372773a0.

91. Berg RJ, van Kranen HJ, Rebel HG, et al. Early p53 alterations in mouse skin carcinogenesis by UVB radiation: immunohistochemical detection of mutant p53 protein in clusters of preneoplastic epidermal cells. Proc Natl Acad Sci U S A. 1996;93:274–8.

92. Greenblatt MS, Bennett WP, Hollstein M, Harris CC. Mutations in the p53 tumor suppressor gene: clues to cancer etiology and molecular pathogenesis. Cancer Res. 1994;54:4855–78.

93. Donehower LA, Harvey M, Slagle BL, et al. Mice deficient for p53 are developmentally normal but susceptible to spontaneous tumours. Nature. 1992;356:215–21. doi:10.1038/356215a0.

94. Jiang W, Ananthaswamy HN, Muller HK, Kripke ML. p53 protects against skin cancer induction by UV-B radiation. Oncogene. 1999;18:4247–53. doi:10.1038/sj.onc.1202789.

95. Amakye D, Jagani Z, Dorsch M. Unraveling the therapeutic potential of the hedgehog pathway in cancer. Nat Med. 2013;19:1410–22. doi:10.1038/nm.3389.

96. Blanpain C, Fuchs E. Epidermal homeostasis: a balancing act of stem cells in the skin. Nat Rev Mol Cell Biol. 2009;10:207–17. doi:10.1038/nrm2636.

97. Burness CB. Sonidegib: first global approval. Drugs. 2015;75:1559–66. doi:10.1007/s40265-015-0458-y.

98. Rudin CM. Vismodegib. Clin Cancer Res. 2012;18:3218–22. doi:10.1158/1078-0432.CCR-12-0568.

99. Gorlin RJ. Nevoid basal cell carcinoma syndrome. Dermatol Clin. 1995;13:113–25.

100. Gorlin RJ, Goltz RW. Multiple nevoid basal-cell epithelioma, jaw cysts and bifid rib. A syndrome. N Engl J Med. 1960;262:908–12. doi:10.1056/NEJM196005052621803.

101. Hahn H, Wicking C, Zaphiropoulous PG, et al. Mutations of the human homolog of Drosophila

patched in the nevoid basal cell carcinoma syndrome. Cell. 1996;85:841–51.

102. Scales SJ, de Sauvage FJ. Mechanisms of hedgehog pathway activation in cancer and implications for therapy. Trends Pharmacol Sci. 2009;30:303–12. doi:10.1016/j.tips.2009.03.007.

103. Epstein EH. Basal cell carcinomas: attack of the hedgehog. Nat Rev Cancer. 2008;8:743–54. doi:10.1038/nrc2503.

104. Heitzer E, Lassacher A, Quehenberger F, et al. UV fingerprints predominate in the PTCH mutation spectra of basal cell carcinomas independent of clinical phenotype. J Invest Dermatol. 2007;127:2872–81. doi:10.1038/sj.jid.5700923.

105. Daya-Grosjean L, Sarasin A. The role of UV induced lesions in skin carcinogenesis: an overview of oncogene and tumor suppressor gene modifications in xeroderma pigmentosum skin tumors. Mutat Res. 2005;571:43–56. doi:10.1016/j.mrfmmm.2004.11.013.

106. Oro AE. Basal cell carcinomas in mice overexpressing Sonic hedgehog. Science. 1997;276:817–21. doi:10.1126/science.276.5313.817.

107. Athar M, Li C, Kim AL, et al. Sonic hedgehog signaling in basal cell nevus syndrome. Cancer Res. 2014;74:4967–75. doi:10.1158/0008-5472.CAN-14-1666.

108. Gober MD, Bashir HM, Seykora JT. Reconstructing skin cancers using animal models. Cancer Metastasis Rev. 2013;32:123–8. doi:10.1007/s10555-012-9410-8.

109. Grachtchouk M, Mo R, Yu S, et al. Basal cell carcinomas in mice overexpressing Gli2 in skin. Nat Genet. 2000;24:216–7. doi:10.1038/73417.

110. Hutchin ME, Kariapper MST, Grachtchouk M, et al. Sustained hedgehog signaling is required for basal cell carcinoma proliferation and survival: conditional skin tumorigenesis recapitulates the hair growth cycle. Genes Dev. 2005;19:214–23. doi:10.1101/gad.1258705.

111. Nilsson M, Undèn AB, Krause D, et al. Induction of basal cell carcinomas and trichoepitheliomas in mice overexpressing GLI-1. Proc Natl Acad Sci U S A. 2000;97:3438–43. doi:10.1073/pnas.050467397.

112. Xie J, Murone M, Luoh SM, et al. Activating Smoothened mutations in sporadic basal-cell carcinoma. Nature. 1998;391:90–2. doi:10.1038/34201.

113. Aszterbaum M, Epstein J, Oro A, et al. Ultraviolet and ionizing radiation enhance the growth of BCCs and trichoblastomas in patched heterozygous knockout mice. Nat Med. 1999;5:1285–91. doi:10.1038/15242.

114. Soufir N, Molès JP, Vilmer C, et al. P16 UV mutations in human skin epithelial tumors. Oncogene. 1999;18:5477–81. doi:10.1038/sj.onc.1202915.

115. Saridaki Z, Liloglou T, Zafiropoulos A, et al. Mutational analysis of CDKN2A genes in patients with squamous cell carcinoma of the skin. Br J Dermatol. 2003;148:638–48.

116. Pierceall WE, Goldberg LH, Tainsky MA, et al. Ras gene mutation and amplification in human nonmelanoma skin cancers. Mol Carcinog. 1991;4:196–202.

117. Van der Lubbe JL, Rosdorff HJ, Bos JL, Van der Eb AJ. Activation of N-ras induced by ultraviolet irradiation in vitro. Oncogene Res. 1988;3:9–20.

118. Miller AJ, Tsao H. New insights into pigmentary pathways and skin cancer. Br J Dermatol. 2010;162:22–8. doi:10.1111/j.1365-2133.2009.09565.x.

119. Gudbjartsson DF, Sulem P, Stacey SN, et al. ASIP and TYR pigmentation variants associate with cutaneous melanoma and basal cell carcinoma. Nat Genet. 2008;40:886–91. doi:10.1038/ng.161.

120. Stacey SN, Sulem P, Masson G, et al. New common variants affecting susceptibility to basal cell carcinoma. Nat Genet. 2009;41:909–14. doi:10.1038/ng.412.

121. Wei YD, Helleberg H, Rannug U, Rannug A. Rapid and transient induction of CYP1A1 gene expression in human cells by the tryptophan photoproduct 6-formylindolo[3,2-b]carbazole. Chem Biol Interact. 1998;110:39–55.

122. Enan E, Matsumura F. Identification of c-Src as the integral component of the cytosolic ah receptor complex, transducing the signal of 2,3,7,8-tetra chlorodibenzo-p-dioxin (TCDD) through the protein phosphorylation pathway. Biochem Pharmacol. 1996;52:1599–612.

123. Kitagawa D, Tanemura S, Ohata S, et al. Activation of extracellular signal-regulated kinase by ultraviolet is mediated through Src-dependent epidermal growth factor receptor phosphorylation. Its implication in an anti-apoptotic function. J Biol Chem. 2002;277:366–71. doi:10.1074/jbc.M107110200.

124. Köhle C, Gschaidmeier H, Lauth D, et al. 2,3,7,8-Tetrachlorodibenzo-p-dioxin (TCDD)-mediated membrane translocation of c-Src protein kinase in liver WB-F344 cells. Arch Toxicol. 1999;73:152–8.

125. Krutmann J, Morita A, Chung JH. Sun exposure: what molecular photodermatology tells us about its good and bad sides. J Invest Dermatol. 2012;132:976–84. doi:10.1038/jid.2011.394.

126. Fritsche E, Schäfer C, Calles C, et al. Lightening up the UV response by identification of the arylhydrocarbon receptor as a cytoplasmatic target for ultraviolet B radiation. Proc Natl Acad Sci U S A. 2007;104:8851–6. doi:10.1073/pnas.0701764104.

127. Aggarwal BB, Gehlot P. Inflammation and cancer: how friendly is the relationship for cancer patients? Curr Opin Pharmacol. 2009;9:351–69. doi:10.1016/j.coph.2009.06.020.

128. Devary Y, Rosette C, DiDonato JA, Karin M. NF-kappa B activation by ultraviolet light not dependent on a nuclear signal. Science. 1993;261:1442–5.

129. Simon MM, Aragane Y, Schwarz A, et al. UVB light induces nuclear factor kappaB (NFkappaB) activity independently from chromosomal DNA damage in cell-free cytosolic extracts. J Invest

Dermatol. 1994;102:422–7. doi:10.1111/1523-1747.ep12372194.

130. Vile GF, Tanew-Ilitschew A, Tyrrell RM. Activation of NF-kappa B in human skin fibroblasts by the oxidative stress generated by UVA radiation. Photochem Photobiol. 1995;62:463–8.

131. Takeda K, Kaisho T, Akira S. Toll-like receptors. Annu Rev Immunol. 2003;21:335–76. doi:10.1146/annurev.immunol.21.120601.141126.

132. Mills KHG. TLR-dependent T cell activation in autoimmunity. Nat Rev Immunol. 2011;11:807–22. doi:10.1038/nri3095.

133. Byrd-Leifer CA, Block EF, Takeda K, et al. The role of MyD88 and TLR4 in the LPS-mimetic activity of Taxol. Eur J Immunol. 2001;31:2448–57. doi:10.1002/1521-4141(200108)31:8<2448::AID-IMMU2448>3.0.CO;2-N.

134. Okamura Y, Watari M, Jerud ES, et al. The extra domain a of fibronectin activates toll-like receptor 4. J Biol Chem. 2001;276:10229–33. doi:10.1074/jbc.M100099200.

135. Termeer C, Benedix F, Sleeman J, et al. Oligosaccharides of Hyaluronan activate dendritic cells via toll-like receptor 4. J Exp Med. 2002;195:99–111.

136. Ohashi K, Burkart V, FLohe S, Kolb H. Cutting edge: heat shock protein 60 is a putative endogenous ligand of the toll-like receptor-4 complex. J Immunol. 2000;164(2):558–61.

137. Kwon M-J, Han J, Kim BH, et al. Superoxide dismutase 3 suppresses hyaluronic acid fragments mediated skin inflammation by inhibition of toll-like receptor 4 signaling pathway: superoxide dismutase 3 inhibits reactive oxygen species-induced trafficking of toll-like receptor 4 to lipid rafts. Antioxid Redox Signal. 2012;16:297–313. doi:10.1089/ars.2011.4066.

138. Kurimoto I, Streilein JW. Characterization of the immunogenetic basis of ultraviolet-B light effects on contact hypersensitivity induction. Immunology. 1994;81:352–8.

139. Lewis W, Simanyi E, Li H, et al. Regulation of ultraviolet radiation induced cutaneous photoimmunosuppression by toll-like receptor-4. Arch Biochem Biophys. 2011;508:171–7. doi:10.1016/j.abb.2011.01.005.

140. Yoshikawa T, Rae V, Bruins-Slot W, et al. Susceptibility to effects of UVB radiation on induction of contact hypersensitivity as a risk factor for skin cancer in humans. J Invest Dermatol. 1990;95:530–6.

141. Ahmad I, Simanyi E, Guroji P, Tamimi IA, delaRosa HJ, Nagar A, Nagar P, Katiyar SK, Elmets CA, Yusuf N. Toll-like receptor-4 deficiency enhances repair of UVR-induced cutaneous DNA damage by nucleotide excision repair mechanism. J Invest Dermatol. 2014;134(6):1710–7.

142. Gao J, Li J, Ma L. Regulation of EGF-induced ERK/MAPK activation and EGFR internalization by G protein-coupled receptor kinase 2. Acta Biochim Biophys Sin. 2005;37:525–31.

143. Pastore S, Mascia F, Mariotti F, et al. ERK1/2 regulates epidermal chemokine expression and skin inflammation. J Immunol. 2005;174:5047–56.

144. Roux PP, Blenis J. ERK and p38 MAPK-activated protein kinases: a family of protein kinases with diverse biological functions. Microbiol Mol Biol Rev. 2004;68:320–44. doi:10.1128/MMBR.68.2.320-344.2004.

145. Wada T, Penninger JM. Mitogen-activated protein kinases in apoptosis regulation. Oncogene. 2004;23:2838–49. doi:10.1038/sj.onc.1207556.

146. Shaulian E, Karin M. AP-1 as a regulator of cell life and death. Nat Cell Biol. 2002;4:E131–6. doi:10.1038/ncb0502-e131.

147. Rosette C, Karin M. Ultraviolet light and osmotic stress: activation of the JNK cascade through multiple growth factor and cytokine receptors. Science. 1996;274:1194–7.

148. Chouinard N, Valerie K, Rouabhia M, Huot J. UVB-mediated activation of p38 mitogen-activated protein kinase enhances resistance of normal human keratinocytes to apoptosis by stabilizing cytoplasmic p53. Biochem J. 2002;365:133–45. doi:10.1042/BJ20020072.

149. Muthusamy V, Piva TJ. The UV response of the skin: a review of the MAPK, NFkappaB and TNFalpha signal transduction pathways. Arch Dermatol Res. 2010;302:5–17. doi:10.1007/s00403-009-0994-y.

150. Assefa Z, Garmyn M, Bouillon R, et al. Differential stimulation of ERK and JNK activities by ultraviolet B irradiation and epidermal growth factor in human keratinocytes. J Invest Dermatol. 1997;108:886–91.

151. An KP, Athar M, Tang X, et al. Cyclooxygenase-2 expression in murine and human nonmelanoma skin cancers: implications for therapeutic approaches. Photochem Photobiol. 2002;76:73–80.

152. Rundhaug JE, Fischer SM. Cyclo-oxygenase-2 plays a critical role in UV-induced skin carcinogenesis. Photochem Photobiol. 2008;84:322–9. doi:10.1111/j.1751-1097.2007.00261.x.

153. Tang X, Kim AL, Kopelovich L, et al. Cyclooxygenase-2 inhibitor nimesulide blocks ultraviolet B-induced photocarcinogenesis in SKH-1 hairless mice. Photochem Photobiol. 2008;84:522–7. doi:10.1111/j.1751-1097.2008.00303.x.

154. Thiagalingam S, editor. Systems biology of cancer. Cambridge: Cambridge University Press; 2015.

155. Bachelor MA, Cooper SJ, Sikorski ET, Bowden GT. Inhibition of p38 mitogen-activated protein kinase and phosphatidylinositol 3-kinase decreases UVB-induced activator protein-1 and cyclooxygenase-2 in a SKH-1 hairless mouse model. Mol Cancer Res MCR. 2005;3:90–9. doi:10.1158/1541-7786.MCR-04-0065.

156. Chen W, Bowden GT. Activation of p38 MAP kinase and ERK are required for ultraviolet-B induced c-fos gene expression in human keratinocytes. Oncogene. 1999;18:7469–76. doi:10.1038/sj.onc.1203210.

157. Tang Q, Gonzales M, Inoue H, Bowden GT. Roles of Akt and glycogen synthase kinase 3beta in the ultraviolet B induction of cyclooxygenase-2 transcription in human keratinocytes. Cancer Res. 2001;61:4329–32.

158. Mahns A, Wolber R, Stäb F, et al. Contribution of UVB and UVA to UV-dependent stimulation of cyclooxygenase-2 expression in artificial epidermis. Photochem Photobiol Sci. 2004;3:257–62. doi:10.1039/b309067a.

159. Santos AL, Oliveira V, Baptista I, et al. Wavelength dependence of biological damage induced by UV radiation on bacteria. Arch Microbiol. 2013;195:63–74. doi:10.1007/s00203-012-0847-5.

160. Gilmour SK. Polyamines and nonmelanoma skin cancer. Toxicol Appl Pharmacol. 2007;224(3):249–56.

161. Hillebrand GG, Winslow MS, Benzinger MJ, Heitmeyer DA, Bissett DL. Acute and chronic ultraviolet radiation induction of epidermal ornithine decarboxylase activity in hairless mice. Cancer Res. 1990;50(5):1580–4.

162. Tang X, Kim AL, Feith DJ, Pegg AE, Russo J, Zhang H, Aszterbaum M, Kopelovich L, Epstein EH Jr, Bickers DR, Athar M. Ornithine decarboxylase is a target for chemoprevention of basal and squamous cell carcinomas in Ptch1+/- mice. J Clin Invest. 2004;113(6):867–75.

163. Bailey HH, Kim K, Verma AK, et al. A randomized, double-blind, placebo-controlled phase 3 skin cancer prevention study of {alpha}-difluoromethylornithine in subjects with previous history of skin cancer. Cancer Prev Res (Phila). 2010;3:35–47. doi:10.1158/1940-6207.CAPR-09-0096.

164. Elmets CA, Cala CM, Xu H. Photoimmunology. Dermatol Clin. 2014;32:277–290., vii. doi:10.1016/j.det.2014.03.005.

165. Ullrich SE, Byrne SN. The immunologic revolution: photoimmunology. J Invest Dermatol. 2012;132:896–905. doi:10.1038/jid.2011.405.

166. Kripke ML. Antigenicity of murine skin tumors induced by ultraviolet light. J Natl Cancer Inst. 1974;53:1333–6.

167. Fisher MS, Kripke ML. Systemic alteration induced in mice by ultraviolet light irradiation and its relationship to ultraviolet carcinogenesis. Proc Natl Acad Sci U S A. 1977;74:1688–92.

168. Sreevidya CS, Fukunaga A, Khaskhely NM, et al. Agents that reverse UV-induced immune suppression and photocarcinogenesis affect DNA repair. J Invest Dermatol. 2010;130:1428–37. doi:10.1038/jid.2009.329.

169. Kripke ML, Fidler IJ. Enhanced experimental metastasis of ultraviolet light-induced fibrosarcomas in ultraviolet light-irradiated syngeneic mice. Cancer Res. 1980;40:625–9.

170. Elmets CA, Bergstresser PR, Tigelaar RE, et al. Analysis of the mechanism of unresponsiveness produced by haptens painted on skin exposed to low dose ultraviolet radiation. J Exp Med. 1983;158:781–94.

171. Greene MI, Sy MS, Kripke M, Benacerraf B. Impairment of antigen-presenting cell function by ultraviolet radiation. Proc Natl Acad Sci U S A. 1979;76:6591–5.

172. Haniszko J, Suskind RR. The effect of ultraviolet radiation on experimental cutaneous sensitization in guinea pigs. J Invest Dermatol. 1963;40:183–91.

173. Jessup JM, Hanna N, Palaszynski E, Kripke ML. Mechanisms of depressed reactivity to dinitrochlorobenzene and ultraviolet-induced tumors during ultraviolet carcinogenesis in BALB/c mice. Cell Immunol. 1978;38:105–15. doi:10.1016/0008-8749(78)90036-9.

174. Morison WL, Parrish JA, Woehler ME, et al. Influence of PUVA and UVB radiation on delayed hypersensitivity in the guinea pig. J Invest Dermatol. 1981;76:484–8.

175. Morison WL, Pike RA, Kripke ML. Effect of sunlight and its component wavebands on contact hypersensitivity in mice and guinea pigs. Photo-Dermatology. 1985;2:195–204.

176. Toews GB, Bergstresser PR, Streilein JW. Epidermal Langerhans cell density determines whether contact hypersensitivity or unresponsiveness follows skin painting with DNFB. J Immunol. 1980;124:445–53.

177. Fukunaga A, Khaskhely NM, Sreevidya CS, et al. Dermal dendritic cells, and not Langerhans cells, play an essential role in inducing an immune response. J Immunol. 2008;180:3057–64. doi:10.4049/jimmunol.180.5.3057.

178. Kaplan DH. In vivo function of Langerhans cells and dermal dendritic cells. Trends Immunol. 2010;31:446–51. doi:10.1016/j.it.2010.08.006.

179. Mathers AR, Larregina AT. Professional antigen-presenting cells of the skin. Immunol Res. 2006;36:127–36. doi:10.1385/IR:36:1:127.

180. Fisher MS, Kripke ML. Suppressor T lymphocytes control the development of primary skin cancers in ultraviolet-irradiated mice. Science. 1982;216:1133–4.

181. Fisher MS, Kripke ML. Further studies on the tumor-specific suppressor cells induced by ultraviolet radiation. J Immunol. 1978;121:1139–44.

182. Kripke ML, Thorn RM, Lill PH, et al. Further characterization of immunological unresponsiveness induced in mice by ultraviolet radiation. Growth and induction of nonultraviolet-induced tumors in ultraviolet-irradiated mice. Transplantation. 1979;28:212–7.

183. Lehtimäki S, Lahesmaa R. Regulatory T cells control immune responses through their non-redundant tissue specific features. Front Immunol. 2013; doi:10.3389/fimmu.2013.00294.

184. Fukunaga A, Khaskhely NM, Ma Y, et al. Langerhans cells serve as immunoregulatory cells by activating NKT cells. J Immunol. 2010;185:4633–40. doi:10.4049/jimmunol.1000246.

185. Loser K, Mehling A, Loeser S, et al. Epidermal RANKL controls regulatory T-cell numbers via acti-

vation of dendritic cells. Nat Med. 2006;12:1372–9. doi:10.1038/nm1518.

186. Aubin F. Mechanisms involved in ultraviolet light-induced immunosuppression. Eur J Dermatol EJD. 2003;13:515–23.

187. Clydesdale GJ, Dandie GW, Muller HK. Ultraviolet light induced injury: immunological and inflammatory effects. Immunol Cell Biol. 2001;79:547–68. doi:10.1046/j.1440-1711.2001.01047.x.

188. Cooper KD, Oberhelman L, Hamilton TA, et al. UV exposure reduces immunization rates and promotes tolerance to epicutaneous antigens in humans: relationship to dose, CD1a-DR+ epidermal macrophage induction, and Langerhans cell depletion. Proc Natl Acad Sci U S A. 1992;89:8497–501.

189. Kang K, Hammerberg C, Meunier L, Cooper KD. CD11b+ macrophages that infiltrate human epidermis after in vivo ultraviolet exposure potently produce IL-10 and represent the major secretory source of epidermal IL-10 protein. J Immunol. 1994;153:5256–64.

190. Moodycliffe AM, Nghiem D, Clydesdale G, Ullrich SE. Immune suppression and skin cancer development: regulation by NKT cells. Nat Immunol. 2000;1:521–5. doi:10.1038/82782.

191. Starcher B. Role for tumour necrosis factor-alpha receptors in ultraviolet-induced skin tumours. Br J Dermatol. 2000;142:1140–7.

192. Bernard JJ, Cowing-Zitron C, Nakatsuji T, et al. Ultraviolet radiation damages self noncoding RNA and is detected by TLR3. Nat Med. 2012;18:1286–90. doi:10.1038/nm.2861.

193. Elmets CA, LeVine MJ, Bickers DR. Action spectrum studies for induction of immunologic unresponsiveness to dinitrofluorobenzene following in vivo low dose ultraviolet radiation. Photochem Photobiol. 1985;42:391–7. doi:10.1111/j.1751-1097.1985.tb01586.x.

194. Kripke ML, Cox PA, Alas LG, Yarosh DB. Pyrimidine dimers in DNA initiate systemic immunosuppression in UV-irradiated mice. Proc Natl Acad Sci U S A. 1992;89:7516–20.

195. Majewski S, Jantschitsch C, Maeda A, et al. IL-23 antagonizes UVR-induced immunosuppression through two mechanisms: reduction of UVR-induced DNA damage and inhibition of UVR-induced regulatory T cells. J Invest Dermatol. 2010;130:554–62. doi:10.1038/jid.2009.274.

196. Schmitt DA, Owen-Schaub L, Ullrich SE. Effect of IL-12 on immune suppression and suppressor cell induction by ultraviolet radiation. J Immunol. 1995;154:5114–20.

197. Schwarz A, Grabbe S, Aragane Y, et al. Interleukin-12 prevents ultraviolet B-induced local immunosuppression and overcomes UVB-induced tolerance. J Invest Dermatol. 1996;106:1187–91.

198. Schwarz A, Maeda A, Ständer S, et al. IL-18 reduces ultraviolet radiation-induced DNA damage and thereby affects photoimmunosuppression. J Immunol. 2006;176:2896–901.

199. Schwarz A, Ständer S, Berneburg M, et al. Interleukin-12 suppresses ultraviolet radiation-induced apoptosis by inducing DNA repair. Nat Cell Biol. 2002;4:26–31. doi:10.1038/ncb717.

200. Schwarz T. 25 years of UV-induced immunosuppression mediated by T cells-from disregarded T suppressor cells to highly respected regulatory T cells. Photochem Photobiol. 2008;84:10–8. doi:10.1111/j.1751-1097.2007.00223.x.

201. Gaspari AA, Fleisher TA, Kraemer KH. Impaired interferon production and natural killer cell activation in patients with the skin cancer-prone disorder, xeroderma pigmentosum. J Clin Invest. 1993;92:1135–42. doi:10.1172/JCI116682.

202. Walterscheid JP, Nghiem DX, Kazimi N, et al. Cis-urocanic acid, a sunlight-induced immunosuppressive factor, activates immune suppression via the 5-HT2A receptor. Proc Natl Acad Sci U S A. 2006;103:17420–5. doi:10.1073/pnas.0603119103.

203. Silverberg MJ, Leyden W, Warton EM, et al. HIV infection status, immunodeficiency, and the incidence of non-melanoma skin cancer. J Natl Cancer Inst. 2013;105:350–60. doi:10.1093/jnci/djs529.

204. Euvrard S, Kanitakis J, Claudy A. Skin cancers after organ transplantation. N Engl J Med. 2003;348:1681–91. doi:10.1056/NEJMra022137.

205. Martinez J-C, Otley CC, Stasko T, et al. Defining the clinical course of metastatic skin cancer in organ transplant recipients: a multicenter collaborative study. Arch Dermatol. 2003;139:301–6.

206. Rowe DE, Carroll RJ, Day CL. Prognostic factors for local recurrence, metastasis, and survival rates in squamous cell carcinoma of the skin, ear, and lip. J Am Acad Dermatol. 1992;26:976–90. doi:10.1016/0190-9622(92)70144-5.

207. Kelly GE, Meikle W, Sheil AG. Scheduled and unscheduled DNA synthesis in epidermal cells of hairless mice treated with immunosuppressive drugs and UVB-UVA irradiation. Br J Dermatol. 1987;117:429–40.

208. Yarosh DB, Pena AV, Nay SL, et al. Calcineurin inhibitors decrease DNA repair and apoptosis in human keratinocytes following ultraviolet B irradiation. J Invest Dermatol. 2005;125:1020–5. doi:10.1111/j.0022-202X.2005.23858.x.

209. Adami J, Frisch M, Yuen J, et al. Evidence of an association between non-Hodgkin's lymphoma and skin cancer. BMJ. 1995;310:1491–5. doi:10.1136/bmj.310.6993.1491.

210. Weimar VM, Ceilley RI, Goeken JA. Cell-mediated immunity in patients with basal and squamous cell skin cancer. J Am Acad Dermatol. 1980;2:143–7.

211. Kaporis HG, Guttman-Yassky E, Lowes MA, et al. Human basal cell carcinoma is associated with Foxp3+ T cells in a Th2 dominant microenvironment. J Invest Dermatol. 2007;127:2391–8. doi:10.1038/sj.jid.5700884.

212. Volden G, Molin L, Thomsen K. PUVA-induced suppression of contact sensitivity to mustine

hydrochloride in mycosis fungoides. Br Med J. 1978;2:865–6.

213. Morison WL, Wimberly J, Parrish JA, Bloch KJ. Abnormal lymphocyte function following long-term PUVA therapy for psoriasis. Br J Dermatol. 1983;108:445–50.

214. Moscicki RA, Morison WL, Parrish JA, et al. Reduction of the fraction of circulating helper-inducer T cells identified by monoclonal antibodies in psoriatic patients treated with long-term psoralen/ultraviolet-a radiation (PUVA). J Invest Dermatol. 1982;79:205–8.

215. Ad Hoc Task Force, Connolly SM, Baker DR, et al. AAD/ACMS/ASDSA/ASMS 2012 appropriate use criteria for Mohs micrographic surgery: a report of the American Academy of Dermatology, American College of Mohs Surgery, American Society for Dermatologic Surgery Association, and the American Society for Mohs Surgery. J Am Acad Dermatol. 2012;67:531–50. doi:10.1016/j.jaad.2012.06.009.

216. Cancer Council Australia, Australian Cancer Network. Clinical practice guide: basal cell carcinoma, squamous cell carcinoma (and related lesions): a guide to clinical management in Australia. Sydney, N.S.W: Cancer Council Australia; 2008.

217. Gupta AK, Paquet M, Villanueva E, Brintnell W. Interventions for actinic keratoses. In: The Cochrane collaboration, editor. Cochrane database of systematic reviews. Chichester: John Wiley & Sons, Ltd; 2012.

218. Miller SJ, Alam M, Andersen J, et al. Basal cell and squamous cell skin cancers. J Natl Compr Cancer Netw. 2010b;8:836–64.

219. Stasko T, Brown MD, Carucci JA, et al. Guidelines for the Management of Squamous Cell Carcinoma in organ transplant recipients. Dermatol Surg. 2004;30:642–50. doi:10.1111/j.1524-4725.2004.30150.x.

220. Stockfleth E, Ferrandiz C, Grob JJ, et al. Development of a treatment algorithm for actinic keratoses: a European consensus. Eur J Dermatol. 2008;18:651–9. doi:10.1684/ejd.2008.0514.

221. Szeimies R-M, Bichel J, Ortonne J-P, et al. A phase II dose-ranging study of topical resiquimod to treat actinic keratosis. Br J Dermatol. 2008;159:205–10. doi:10.1111/j.1365-2133.2008.08615.x.

222. Zhang G, Dass CR, Sumithran E, et al. Effect of deoxyribozymes targeting c-Jun on solid tumor growth and angiogenesis in rodents. J Natl Cancer Inst. 2004;96:683–96.

223. Cho E-A, Moloney FJ, Cai H, et al. Safety and tolerability of an intratumorally injected DNAzyme, Dz13, in patients with nodular basal-cell carcinoma: a phase 1 first-in-human trial (DISCOVER). Lancet. 2013;381:1835–43. doi:10.1016/S0140-6736(12)62166-7.

224. Kim J, Tang JY, Gong R, et al. Itraconazole, a commonly used antifungal that inhibits hedgehog pathway activity and cancer growth. Cancer Cell. 2010;17:388–99. doi:10.1016/j.ccr.2010.02.027.

225. Kim DJ, Kim J, Spaunhurst K, et al. Open-label, exploratory phase II trial of oral itraconazole for the treatment of basal cell carcinoma. J Clin Oncol. 2014;32:745–51. doi:10.1200/JCO.2013.49.9525.

226. Bauman JE, Eaton KD, Martins RG. Treatment of recurrent squamous cell carcinoma of the skin with cetuximab. Arch Dermatol. 2007;143:889–92. doi:10.1001/archderm.143.7.889.

227. Maubec E, Petrow P, Scheer-Senyarich I, et al. Phase II study of cetuximab as first-line single-drug therapy in patients with unresectable squamous cell carcinoma of the skin. J Clin Oncol. 2011;29:3419–26. doi:10.1200/JCO.2010.34.1735.

228. Giacchero D, Barrière J, Benezery K, et al. Efficacy of cetuximab for unresectable or advanced cutaneous squamous cell carcinoma--a report of eight cases. Clin Oncol (R Coll Radiol). 2011;23:716–8. doi:10.1016/j.clon.2011.07.007.

229. Eder J, Simonitsch-Klupp I, Trautinger F. Treatment of unresectable squamous cell carcinoma of the skin with epidermal growth factor receptor antibodies-a case series. Eur J Dermatol. 2013;23:658–62. doi:10.1684/ejd.2013.2153.

230. Foote MC, McGrath M, Guminski A, et al. Phase II study of single-agent panitumumab in patients with incurable cutaneous squamous cell carcinoma. Ann Oncol. 2014;25:2047–52. doi:10.1093/annonc/mdu368.

231. Lewis CM, Glisson BS, Feng L, et al. A phase II study of gefitinib for aggressive cutaneous squamous cell carcinoma of the head and neck. Clin Cancer Res. 2012;18:1435–46. doi:10.1158/1078-0432.CCR-11-1951.

232. Khan N, Afaq F, Mukhtar H. Cancer chemoprevention through dietary antioxidants: progress and promise. Antioxid Redox Signal. 2008;10:475–510. doi:10.1089/ars.2007.1740.

233. Sambandan DR, Ratner D. Sunscreens: an overview and update. J Am Acad Dermatol. 2011;64:748–58. doi:10.1016/j.jaad.2010.01.005.

234. Green A, Williams G, Neale R, et al. Daily sunscreen application and betacarotene supplementation in prevention of basal-cell and squamous-cell carcinomas of the skin: a randomised controlled trial. Lancet. 1999;354:723–9. doi:10.1016/S0140-6736(98)12168-2.

235. Naylor MF, Boyd A, Smith DW, et al. High sun protection factor sunscreens in the suppression of actinic neoplasia. Arch Dermatol. 1995;131:170–5.

236. Thompson SC, Jolley D, Marks R. Reduction of solar keratoses by regular sunscreen use. N Engl J Med. 1993;329:1147–51. doi:10.1056/NEJM199310143291602.

237. Ulrich C, Jürgensen JS, Degen A, et al. Prevention of non-melanoma skin cancer in organ transplant patients by regular use of a sunscreen: a 24 months, prospective, case-control study.

Br J Dermatol. 2009;161(Suppl 3):78–84. doi:10.1111/j.1365-2133.2009.09453.x.

238. van der Pols JC, Williams GM, Pandeya N, et al. Prolonged prevention of squamous cell carcinoma of the skin by regular sunscreen use. Cancer Epidemiol Biomarkers Prev. 2006;15:2546–8. doi:10.1158/1055-9965.EPI-06-0352.

239. Harvey I, Frankel S, Marks R, et al. Non-melanoma skin cancer and solar keratoses II analytical results of the South Wales skin cancer study. Br J Cancer. 1996;74:1308–12.

240. Handel AE, Ramagopalan SV. The questionable effectiveness of sunscreen. Lancet. 2010;376:161–162.; author reply 162. doi:10.1016/S0140-6736(10)61104-X.

241. Hood WF, Gierse JK, Isakson PC, et al. Characterization of celecoxib and valdecoxib binding to cyclooxygenase. Mol Pharmacol. 2003;63:870–7.

242. Elmets CA, Viner JL, Pentland AP, et al. Chemoprevention of nonmelanoma skin cancer with celecoxib: a randomized, double-blind, placebo-controlled trial. J Natl Cancer Inst. 2010;102:1835–44. doi:10.1093/jnci/djq442.

243. Weinstock MA, Bingham SF, Lew RA, et al. Topical tretinoin therapy and all-cause mortality. Arch Dermatol. 2009;145:18–24. doi:10.1001/archdermatol.2008.542.

244. Kadakia KC, Barton DL, Loprinzi CL, et al. Randomized controlled trial of acitretin versus placebo in patients at high-risk for basal cell or squamous cell carcinoma of the skin (north central cancer treatment group study 969251). Cancer. 2012;118:2128–37. doi:10.1002/cncr.26374.

245. Cafardi JA, Shafi R, Athar M, Elmets CA. Prospects for skin cancer treatment and prevention: the potential contribution of an engineered virus. J Invest Dermatol. 2011;131:559–61. doi:10.1038/jid.2010.394.

246. DeBoyes T, Kouba D, Ozog D, et al. Reduced number of actinic keratoses with topical application of DNA repair enzyme creams. J Drugs Dermatol. 2010;9:1519–21.

247. Yarosh D, Bucana C, Cox P, et al. Localization of liposomes containing a DNA repair enzyme in murine skin. J Invest Dermatol. 1994;103:461–8. doi:10.1111/1523-1747.ep12395551.

248. Yarosh D, Klein J, O'Connor A, et al. Effect of topically applied T4 endonuclease V in liposomes on skin cancer in xeroderma pigmentosum: a randomised study. Xeroderma Pigmentosum Study Group. Lancet. 2001;357:926–9.

249. Zahid S, Brownell I. Repairing DNA damage in xeroderma pigmentosum: T4N5 lotion and gene therapy. J Drugs Dermatol. 2008;7:405–8.

250. Jang M, Cai L, Udeani GO, et al. Cancer chemopreventive activity of resveratrol, a natural product derived from grapes. Science. 1997;275:218–20.

251. Meeran SM, Mantena SK, Meleth S, et al. Interleukin-12-deficient mice are at greater risk of UV radiation-induced skin tumors and malignant transformation of papillomas to carcinomas. Mol Cancer Ther. 2006;5:825–32. doi:10.1158/1535-7163.MCT-06-0003.

252. Mittal A, Elmets CA, Katiyar SK. Dietary feeding of proanthocyanidins from grape seeds prevents photocarcinogenesis in SKH-1 hairless mice: relationship to decreased fat and lipid peroxidation. Carcinogenesis. 2003;24:1379–88. doi:10.1093/carcin/bgg095.

253. Roy AM, Baliga MS, Elmets CA, Katiyar SK. Grape seed Proanthocyanidins induce apoptosis through p53, Bax, and caspase 3 pathways. Neoplasia. 2005;7:24–36. doi:10.1593/neo.04412.

254. Sharma SD, Katiyar SK. Dietary grape-seed proanthocyanidin inhibition of ultraviolet B-induced immune suppression is associated with induction of IL-12. Carcinogenesis. 2006;27:95–102. doi:10.1093/carcin/bgi169.

255. Yang Y, Paik JH, Cho D, et al. Resveratrol induces the suppression of tumor-derived CD4+CD25+ regulatory T cells. Int Immunopharmacol. 2008;8:542–7. doi:10.1016/j.intimp.2007.12.006.

256. Yusuf N, Nasti TH, Meleth S, Elmets CA. Resveratrol enhances cell-mediated immune response to DMBA through TLR4 and prevents DMBA induced cutaneous carcinogenesis. Mol Carcinog. 2009;48:713–23. doi:10.1002/mc.20517.

257. Katiyar S, Elmets CA, Katiyar SK. Green tea and skin cancer: photoimmunology, angiogenesis and DNA repair. J Nutr Biochem. 2007;18:287–96. doi:10.1016/j.jnutbio.2006.08.004.

258. Hora JJ, Maydew ER, Lansky EP, Dwivedi C. Chemopreventive effects of pomegranate seed oil on skin tumor development in CD1 mice. J Med Food. 2003;6:157–61. doi:10.1089/10966200360716553.

259. Burton A. Chemoprevention: eat ginger, rub on pomegranate. Lancet Oncol. 2003;4:715.

260. Afaq F, Saleem M, Krueger CG, et al. Anthocyanin- and hydrolyzable tannin-rich pomegranate fruit extract modulates MAPK and NF-kappaB pathways and inhibits skin tumorigenesis in CD-1 mice. Int J Cancer. 2005b;113:423–33. doi:10.1002/ijc.20587.

261. Afaq F, Malik A, Syed D, et al. Pomegranate fruit extract modulates UV-B-mediated phosphorylation of mitogen-activated protein kinases and activation of nuclear factor kappa B in normal human epidermal keratinocytes paragraph sign. Photochem Photobiol. 2005a;81:38–45. doi:10.1562/2004-08-06-RA-264.

262. Afaq F, Zaid MA, Khan N, et al. Protective effect of pomegranate-derived products on UVB-mediated damage in human reconstituted skin. Exp Dermatol. 2009;18:553–61. doi:10.1111/j.1600-0625.2008.00829.x.

263. Afaq F, Khan N, Syed DN, Mukhtar H. Oral feeding of pomegranate fruit extract inhibits early biomarkers of UVB radiation-induced carcinogenesis in SKH-1 hairless mouse epidermis. Photochem Photobiol. 2010;86:1318–26. doi:10.1111/j.1751-1097.2010.00815.x.

264. de Vries E, Trakatelli M, Kalabalikis D, et al. Known and potential new risk factors for skin cancer in European populations: a multicentre case-control study. Br J Dermatol. 2012;167(Suppl 2):1–13. doi:10.1111/j.1365-2133.2012.11081.x.

265. Gensler HL, Williams T, Huang AC, Jacobson EL. Oral niacin prevents photocarcinogenesis and photoimmunosuppression in mice. Nutr Cancer. 1999;34:36–41. doi:10.1207/S15327914NC340105.

266. Schreiber V, Dantzer F, Ame J-C, de Murcia G. Poly(ADP-ribose): novel functions for an old molecule. Nat Rev Mol Cell Biol. 2006;7:517–28. doi:10.1038/nrm1963.

267. Fisher AEO, Hochegger H, Takeda S, Caldecott KW. Poly(ADP-ribose) polymerase 1 accelerates single-strand break repair in concert with poly(ADP-ribose) Glycohydrolase. Mol Cell Biol. 2007;27:5597–605. doi:10.1128/MCB.02248-06.

268. Damian DL, Patterson CRS, Stapelberg M, et al. UV radiation-induced immunosuppression is greater in men and prevented by topical nicotinamide. J Invest Dermatol. 2008;128:447–54. doi:10.1038/sj.jid.5701058.

269. Yiasemides E, Sivapirabu G, Halliday GM, et al. Oral nicotinamide protects against ultraviolet radiation-induced immunosuppression in humans. Carcinogenesis. 2009;30:101–5. doi:10.1093/carcin/bgn248.

270. Park J, Halliday GM, Surjana D, Damian DL. Nicotinamide prevents ultraviolet radiation-induced cellular energy loss. Photochem Photobiol. 2010;86:942–8. doi:10.1111/j.1751-1097.2010.00746.x.

271. Surjana D, Halliday GM, Martin AJ, et al. Oral nicotinamide reduces actinic keratoses in phase II double-blinded randomized controlled trials. J Invest Dermatol. 2012;132:1497–500. doi:10.1038/jid.2011.459.

272. Benavente CA, Schnell SA, Jacobson EL. Effects of niacin restriction on Sirtuin and PARP responses to photodamage in human skin. PLoS One. 2012;7:e42276. doi:10.1371/journal.pone.0042276.

273. Martin AJ, Chen A, Penas PF, Halliday G, Dalziell R, McKenzie C, Scolyer RA, Dhillon HM, Vardy JL, George GS, Chinniah N, Damian D. Oral nicotinamide to reduce actinic cancer: a phase 3 double-blind randomized controlled trial. J Clin Oncol. 2015;33(15):9000. doi:10.1200/jco.2015.33.15.

274. Gu M, Dhanalakshmi S, Singh RP, Agarwal R. Dietary feeding of silibinin prevents early biomarkers of UVB radiation-induced carcinogenesis in SKH-1 hairless mouse epidermis. Cancer Epidemiol Biomarkers Prev. 2005;14:1344–9. doi:10.1158/1055-9965.EPI-04-0664.

275. Vaid M, Katiyar SK. Molecular mechanisms of inhibition of photocarcinogenesis by silymarin, a phytochemical from milk thistle (Silybum marianum L. Gaertn.) (review). Int J Oncol. 2010;36(5):1053–60. doi:10.3892/ijo_00000586.

276. Vaid M, Prasad R, Singh T, et al. Silymarin inhibits ultraviolet radiation-induced immune suppression through DNA repair-dependent activation of dendritic cells and stimulation of effector T cells. Biochem Pharmacol. 2013;85:1066–76. doi:10.1016/j.bcp.2013.01.026.

277. Middelkamp-Hup MA, Pathak MA, Parrado C, et al. Oral polypodium leucotomos extract decreases ultraviolet-induced damage of human skin. J Am Acad Dermatol. 2004;51:910–8. doi:10.1016/j.jaad.2004.06.027.

278. Jańczyk A, Garcia-Lopez MA, Fernandez-Peñas P, et al. A Polypodium leucotomos extract inhibits solar-simulated radiation-induced TNF-alpha and iNOS expression, transcriptional activation and apoptosis. Exp Dermatol. 2007;16:823–9. doi:10.1111/j.1600-0625.2007.00603.x.

279. Philips N, Conte J, Chen Y-J, et al. Beneficial regulation of matrixmetalloproteinases and their inhibitors, fibrillar collagens and transforming growth factor-beta by Polypodium leucotomos, directly or in dermal fibroblasts, ultraviolet radiated fibroblasts, and melanoma cells. Arch Dermatol Res. 2009;301:487–95. doi:10.1007/s00403-009-0950-x.

280. Rodríguez-Yanes E, Juarranz Á, Cuevas J, et al. Polypodium leucotomos decreases UV-induced epidermal cell proliferation and enhances p53 expression and plasma antioxidant capacity in hairless mice. Exp Dermatol. 2012;21:638–40. doi:10.1111/j.1600-0625.2012.01544.x.

281. El-Haj N, Goldstein N. Sun protection in a pill: the photoprotective properties of Polypodium leucotomos extract. Int J Dermatol. 2015;54:362–6. doi:10.1111/ijd.12611.

282. Berman B, Ellis C, Elmets C. Polypodium Leucotomos – an overview of basic investigative findings. J Drugs Dermatol. 2016;15:224–8.

283. Howes LG. Selective COX-2 inhibitors, NSAIDs and cardiovascular events – is celecoxib the safest choice? Ther Clin Risk Manag. 2007;3:831–45.

284. Silverstein FE, Faich G, Goldstein JL, et al. Gastrointestinal toxicity with celecoxib vs nonsteroidal anti-inflammatory drugs for osteoarthritis and rheumatoid arthritis: the CLASS study: a randomized controlled trial. Celecoxib long-term arthritis safety study. JAMA. 2000;284:1247–55.

285. Zhang J, Ding EL, Song Y. Adverse effects of cyclooxygenase 2 inhibitors on renal and arrhythmia events: meta-analysis of randomized trials. JAMA. 2006;296:1619. doi:10.1001/jama.296.13.jrv60015.

286. Bavinck JN, Tieben LM, Van der Woude FJ, et al. Prevention of skin cancer and reduction of keratotic skin lesions during acitretin therapy in renal transplant recipients: a double-blind, placebo-controlled study. J Clin Oncol. 1995;13:1933–8.

287. George R, Weightman W, Russ GR, et al. Acitretin for chemoprevention of non-melanoma skin cancers in renal transplant recipients. Australas J Dermatol. 2002;43:269–73.

288. Kraemer KH, DiGiovanna JJ, Moshell AN, et al. Prevention of skin cancer in xeroderma pigmentosum with the use of oral isotretinoin. N Engl J Med. 1988;318:1633–7. doi:10.1056/NEJM198806233182501.

289. Peck GL, DiGiovanna JJ, Sarnoff DS, et al. Treatment and prevention of basal cell carcinoma with oral isotretinoin. J Am Acad Dermatol. 1988;19:176–85.

290. Tangrea JA, Edwards BK, Taylor PR, et al. Long-term therapy with low-dose isotretinoin for prevention of basal cell carcinoma: a multicenter clinical trial. Isotretinoin-basal cell carcinoma study group. J Natl Cancer Inst. 1992;84:328–32.

291. So P-L, Lee K, Hebert J, et al. Topical tazarotene chemoprevention reduces basal cell carcinoma number and size in Ptch1+/− mice exposed to ultraviolet or ionizing radiation. Cancer Res. 2004;64:4385–9. doi:10.1158/0008-5472.CAN-03-1927.

292. Peris K, Fargnoli MC, Chimenti S. Preliminary observations on the use of topical tazarotene to treat basal-cell carcinoma. N Engl J Med. 1999;341:1767–8. doi:10.1056/NEJM199912023412312.

293. Atigadda VR, Xia G, Desphande A, et al. Methyl substitution of a rexinoid agonist improves potency and reveals site of lipid toxicity. J Med Chem. 2014;57:5370–80. doi:10.1021/jm5004792.

294. Brtko J, Thalhamer J. Renaissance of the biologically active vitamin a derivatives: established and novel directed therapies for cancer and chemoprevention. Curr Pharm Des. 2003;9:2067–77.

295. Desphande A, Xia G, Boerma LJ, et al. Methyl-substituted conformationally constrained rexinoid agonists for the retinoid X receptors demonstrate improved efficacy for cancer therapy and prevention. Bioorg Med Chem. 2014;22:178–85. doi:10.1016/j.bmc.2013.11.039.

296. Chen AC, Martin AJ, Choy B, et al. A phase 3 randomized trial of nicotinamide for skin-cancer chemoprevention. N Engl J Med. 2015;373:1618–26. doi:10.1056/NEJMoa1506197.

Ambient Particulate Matter and Skin

9

Andrea Vierkötter, Jean Krutmann, and Tamara Schikowski

9.1 Ambient Particulate Matter and Its Health Effects

9.1.1 Characteristics of Ambient Particulate Matter

Ambient particulate matter (PM) is one of the components of ambient air pollution, which can be produced either by natural processes such as volcanic activity or dust storms or by human activity such as fossil fuel combustion or chemical production. Furthermore, air pollutants including PM can be classified as either primary or secondary, depending on how they were formed. Primary pollutants are those emitted directly. Secondary pollutants are those that form when primary pollutants react or interact with each other in the atmosphere leading to transformation products. PM pollution is not a specific single component of air pollution, but a complex mixture of extremely small particles and droplets. It consists of a number of components including acids, organic chemicals, metals, and soil or dust particles and is typically categorized by its size (Table 9.1) and penetration capabilities.

A. Vierkötter (✉) • J. Krutmann • T. Schikowski
IUF—Leibniz Institut für umweltmedizinische
Forschung gGmbH, Auf'm Hennekamp 50,
40225 Düsseldorf, Germany
e-mail: Andrea.Vierkoetter@IUF-Duesseldorf.de;
Tamara.Schikowski@IUF-Duesseldorf.de

Beyond others, two main categories according to particle size and penetration capability are PM_{10}, which is the fraction of suspended particles 10 μm in diameter and smaller that is able to enter the nasal cavity and $PM_{2.5}$, which has a maximum particle size of 2.5 μm and is able to enter the bronchia and lungs.

The PM fractions described in Table 9.1 not only reflect the particle size but also its source and formation process. In this regard, coarse particles derive primarily from suspension of dust, soil, or other crustal materials from roads, farming, mining, windstorms, or volcanoes. They also include sea salts, pollen, mold, spores, and other biological materials. Fine particles are mainly derived from direct emissions from combustion processes, such as vehicle use of gasoline and diesel, wood burning, coal burning for power

Table 9.1 Commonly defined particulate matter (PM) fractions according to particle size

PM fraction	Particle size
total suspended particles (TSP)	includes all particles up to 30 μm in diameter
PM_{10}	particles with a diameter ≤ 10 μm
coarse particles	particles with a diameter of 2.5–10 μm
$PM_{2.5}$ or fine particles	particles with a diameter ≤ 2.5 μm
ultrafine particles or $PM_{0.1}$ or nanoparticles (for engineered material)	particles with a diameter ≤ 0.1 μm

© Springer International Publishing Switzerland 2018
J. Krutmann, H.F. Merk (eds.), *Environment and Skin*,
https://doi.org/10.1007/978-3-319-43102-4_9

generation, and industrial processes, such as smelters, cement plants, paper mills, and steel mills. Fine particles also consist of transformation products including nitrate and sulfate particles. Ultrafine particles are typically fresh emissions from combustion-related sources, such as vehicle exhaust and atmospheric photochemical reactions. They have a very short life in the range of minutes or hours as they grow rapidly through coagulation and/or condensation to larger complex aggregates in the $PM_{2.5}$ range.

In Europe a regulatory framework has been established to reduce air pollution including PM emissions. However, ambient concentrations across Europe still exceed the short- and long-term standards set by the European Union. According to the latest air quality directive, the Directive on Ambient Air Quality and Cleaner Air for Europe (Directive 2008/50/EC), the daily limit value of PM_{10} is 50 µg/m³ not to be exceeded on more than 35 days, and the annual average should be not higher than 40 µg/m³. For $PM_{2.5}$ there is no daily limit but the annual average should be not higher than 25 µg/m³ (Table 9.2). However, the European guidelines are even less strict than the recommendations by the World Health Organization (WHO) developed in 2005 (Table 9.2).

The WHO PM limits aim to offer guidance to policy makers in reducing the health impacts of air pollution based on expert evaluation and current scientific evidence. This means that the actual European limits still have detrimental effects on the health status of the population. In other countries the situation is even worse. Worldwide PM levels have been obtained by satellite images, and this revealed that $PM_{2.5}$ levels are particularly high in northern India and China (Fig. 9.1) reaching an annual average over 50 µg/m³ [4].

Table 9.2 Clean air policy targets for PM_{10} and $PM_{2.5}$ set by the World Health Organization (WHO), the European Union (EU) and the USA

Source	PM_{10} [µg/m³]		$PM_{2.5}$ [µg/m³]	
	1 year	24 h	1 year	24 h
WHO [1]	20	50[a]	10	25[a]
EU [2]	40	50[b]	25	
USA [3]	50	150	15	65

[a]Not to be exceeded more than 3 days per year
[b]Not to be exceeded more than 35 days per year

9.1.2 Health Effects of Ambient Particulate Matter

In 2012 the WHO reported that 3.7 million deaths are attributed to ambient air pollution worldwide. This places air pollution as the current world's largest single environmental health risk factor.

Fig. 9.1 Air pollution in Shanghai City (China) in 2014 (Source: Anke Hüls, IUF)

Furthermore, in 2013 the International Agency for Research on Cancer (IARC) has classified outdoor air pollution as well as particulate matter as carcinogenic to humans. In Europe, the USA and China studies repeatedly show that the acute and long-term effects of PM include increased cardiovascular and respiratory morbidity and mortality [5–10]. Additionally, there is growing evidence that other organs might also be affected by ambient particulate pollution. Recent reports indicate a link between particle pollution and the occurrence of neurodegenerative diseases such as cognitive dysfunction, Alzheimer's, and potentially Parkinson's diseases [11, 12]. Recently, there is increasing evidence that ambient particle pollution may exert negative effects on human skin.

9.2 Epidemiological Studies on Effects of Ambient Particulate Matter on Skin

Although epidemiological studies highlight the adverse effects of pollution on human health, very little is known on its adverse effects on the skin. However, there is increasing evidence in the recent years that ambient PM exposure impacts human skin as well. The effect of ambient air pollution on the skin seems to be obvious because the skin is our outermost barrier and thus in direct contact with ambient air pollutants.

9.2.1 Ambient Particulate Matter Exposure and Diseased Skin

Diseased skin with an impact in skin barrier function might be more susceptible to ambient PM exposure than healthy skin as particles or substances bound to the particles might penetrate more easily through diseased skin as through healthy skin. There are several studies, which have shown a link between ambient air pollution and an increased risk of symptoms of eczema. In the following three corresponding studies are presented.

The first study by Larrieu *et al.* [13] explored the link between daily levels of air pollution including PM_{10} and medical home visits made for

diverse reasons including skin rash in Bordeaux (France) in 2000–2006. Skin rash was chosen as a health indicator for dermatitis, eczema, and urticaria. The daily number of home visits was obtained from a network of general practitioners called SOS Médecins, and data on daily levels of ambient air pollutants measured within the study area was provided by the local air quality monitoring network AIRAQ. The study area comprised 22 cities and more than 600,000 inhabitants. During the 7-year study period, a total of 895,710 medical home visits were made by SOS Médecins Bordeaux corresponding to a daily mean of 350 visits. It was found that the risk for the health indicator skin rash was increased by 3.2% (95% CI: −0.2–6.8%) during the 3 days following increases in PM_{10}. A second study investigated the association between traffic-related air pollution ($PM_{2.5}$ absorbance and NO_2) and the incidence and prevalence of respiratory allergies and eczema in children recruited from the lower polluted Wesel area of North Rhine-Westphalia (Germany) [14]. Significant and positive associations between traffic-related air pollution and respiratory allergies and eczema were already shown for the metropolitan area of Munich (Germany) [15], which is highly polluted. In the study performed in the Wesel area, 3390 newborns between 1995 and 1999 were recruited and followed up till the age of six. Diagnoses and symptoms of respiratory allergies and eczema were recorded by annual questionnaires. At each follow-up the parents were asked whether a physician had diagnosed atopic eczema in the child since the last follow-up. They were additionally asked whether the child had experienced an intermittent, itchy skin rash that lasted at least 2 weeks. At the age of six a clinical test for eczema and IgE sensitization was performed. Individual exposure to traffic-related $PM_{2.5}$ absorbance and NO_2 at the children's home addresses was determined by land-use regression. Exposure was further characterized by the distance from the domicile to the next major road. Doctors' diagnosed eczema occurred in more than 20% of the children between birth and age six. Eight percent still had doctor diagnosed eczema in the sixth year of life. The point prevalence at the clinical investigation was 4%. The prevalence of eczema at age six was significantly higher in children who reside in areas with higher traffic-related air pollution. The adjusted relative risk for doctor diagnosed eczema, for instance, was 1.69 (95% CI: 1.04–2.75) per 90% range of soot concentration. Current eczema was likewise associated. Children with parental allergies showed significantly stronger effects. However, incidence of eczema was not associated with traffic-related air pollution. The authors concluded that the effect of traffic-related air pollution was due to the prolongation of eczema not due to the incidence of new cases. In a third study, Kim *et al.* [16] showed that ambient air pollution acts as an aggravating factor for atopic dermatitis (AD). These authors carried out a long-term study in Seoul (Korea) to evaluate the clinical effects of outdoor air pollutants on skin symptoms in children with atopic dermatitis. The study was conducted in 2009 to 2010 over a period of 18 months. The presence of AD symptoms on each day was defined by using a symptom scale of 4 or more for itching and sleep disturbance accompanied by at least one of the following symptoms: erythema, edema or oozing. Air pollution concentrations of the nearest monitoring site to the residential address were assigned to each study participant. Throughout the whole study period, concentrations of different ambient pollutants including PM were higher on days when patients had AD symptoms than on days without AD symptoms. An increase of PM_{10} by 1 $\mu g/m^3$ was significantly associated with a 0.44% (95% confidence interval (CI): 0.12–0.77%) increase in AD symptoms on the following day.

9.2.2   Effects on Healthy Skin

The first indication that ambient particulate matter exposure is also capable of negatively impacting healthy human skin came from a study published by Vierkötter *et al.* [17]. They showed that chronic traffic-related PM exposure is significantly associated with extrinsic skin aging in an epidemiological study. The epidemiological data came from a cohort of women, the SALIA cohort, which was initiated in Germany between 1985 and 1994 as part of the Environmental Health Survey, an element of the clean air plan introduced by the government of North Rhine-Westphalia (NRW) to assess the influence of air pollution exposure on lung diseases. The areas were chosen to represent a range of exposure from ambient PM from traffic, steel, and coal industries. At baseline, all women aged 54–55 living in predefined areas were asked to participate. Follow-up questionnaires were conducted in 2006, and follow-up investigations were performed in 2007–2009 and in 2012 and 2013 including skin aging assessments. Using the SALIA study data, Vierkötter *et al.* [17] showed for the first time the association between ambient PM exposure and extrinsic skin aging, particularly between ambient soot exposure and pigment spot formation. Specifically, the pigment spot occurrence increased 1.2-fold (95% CI: 1.03–1.40) per interquartile range of soot exposure (Fig. 9.2).

Another study has investigated the impact of urban pollution on biochemical and clinical parameters of the skin in volunteers living in Mexico [18]. In total, 189 healthy volunteers (96 volunteers from Mexico City and 93 volunteers from Cuernavaca) were enrolled. Per study design skin parameters of volunteers living in the highly polluted environment of Mexico City were compared to the skin parameters of volunteers living in the less polluted environment of Cuernavaca 50 km away from Mexico City. On the face, skin pH and moisturizing level, sebum production, erythematous, and melanin index were measured. Furthermore, for the measurement of biochemical parameters, sebum was harvested by cotton pads, and superficial stratum corneum was obtained by tape stripping. In sebum samples, the antioxidants squalene and vitamin E were measured as well as lactic acid and cholesterol levels. In samples of the stratum corneum, the pro-inflammatory marker interleukin (IL)-1α, adenosine triphosphate (ATP), corneodesmosin, oxidized proteins, and trypsin-like and chymotrypsin-like activity were measured. In addition, the volunteers were asked

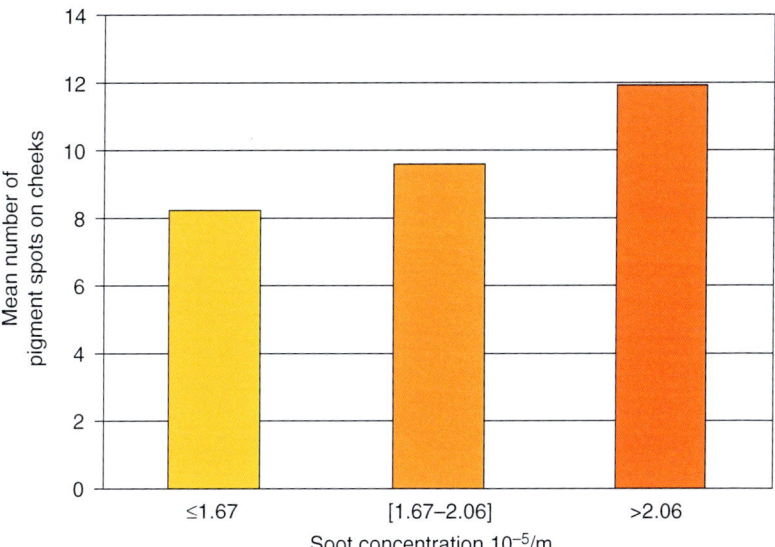

Fig. 9.2 Increase of the mean number of pigment spots on cheeks with increasing soot concentration in the SALIA study cohort

about some clinical symptoms of their skin by dermatologists. The study provided evidence for significant differences between some of the measured skin parameters in the two study areas. For example, the moisturizing level measured was significantly higher in Cuernavaca than in Mexico City indicating a dryer skin in volunteers living in Mexico City. Furthermore, sebum samples of volunteers from Mexico City showed lower squalene and vitamin E levels than sebum samples of volunteers from Cuernavaca. Moreover, oxidized proteins in stratum corneum samples were higher in samples from volunteers from Mexico City, whereas IL-1α and ATP levels were higher in samples from Cuernavaca volunteers. All these results indicate that urban pollution has an impact on skin quality.

9.3 Possible Mechanistic Explanations of Ambient Particulate Matter Induced Skin Effects

9.3.1 Outside-Inside Effect

From a theoretical point of view, ambient particulate matter (PM) might act over an outside-inside effect meaning that PM penetrates directly from the air into the skin. Penetration capability of par-

ticles from the air into the skin is not well investigated and depends on particle size and on the chemical composition (inorganic or organic) of the particle [19, 20]. In the lung PM exposure has been linked to the generation of reactive oxygen species (ROS) [21]. Whether this is also true for skin cells is currently not known. The biological effects observed in the epidemiological studies might also be attributed to substances which are bound to the particles. For example, PM from incomplete combustion carry organic chemicals such as polycyclic aromatic hydrocarbons (PAHs) on its surface. PAHs are highly lipophilic and easily penetrate the skin. As described above, Vierkötter et al. [17] have found the strongest association between extrinsic skin aging and soot exposure, a mixture of carbon particles covered with PAHs. The mode of action of PAHs is most probably the activation of arylhydrocarbon receptor (AhR) signaling in the skin as PAHs are ligands of the AhR and the AhR is expressed in virtually all skin cell populations assessed including keratinocytes and melanocytes [22–24]. There is experimental evidence from human keratinocyte cultures exposed to Asian dust storm particles that this might cause increased AhR expression and upregulation of pro-inflammatory mediators [25]. Indeed, recent own mechanistic studies show that ex vivo exposure of human skin to ambient relevant diesel exhaust affects gene transcription including AhR signature genes such as cytochrome P450 (CYP) 1A1 (Krutmann *et al.*, unpublished). One biological consequence of AhR activation and increased CYP1A1 expression is an increased production of ROS indicating the possibility that topically applied AhR antagonists [24, 26] as well as antioxidants might be useful in protecting human skin against PM-induced detrimental effects. These studies, however, do not exclude the possibilities that also the particle itself (in addition to the surface-bounded material) might have the capacity to influence skin cells as has been shown to be the case for lung epithelial cells.

9.3.2 Inside-Outside Effect

Another possibility can be an inside-outside effect. This means that PM exposure might cause systemic effects as a consequence of particles penetrating the lung and subsequently the circulation or by causing an inflammatory reaction in the lung which may subsequently cause systemic inflammatory reactions. Systemic inflammation might then be detrimental to the skin as well. The same mode of action was already hypothesized for the effect of air pollution on cardiovascular diseases [27], but has to be further investigated regarding its effect on the skin.

9.4 Conclusion and Outlook

Although there are guidelines for the restriction of ambient particle exposure in many countries, the current air pollution levels might still have a negative impact on human health including skin health. Ambient PM does not only affect diseased skin by triggering atopic dermatitis symptoms but also affects intact, healthy skin by influencing physiological skin properties and by contributing to extrinsic skin aging. There might be different mechanisms how ambient PM exposure affects diseased or healthy skin, which have to be further investigated in the future. Furthermore, it will be also important to investigate the influence of the ethnic background on the effect of ambient PM and skin health. Moreover, there might be susceptible

groups in one population referring to their specific genetic makeup which show an increased effect of ambient PM exposure and certain skin outcomes. In conclusion, more research is needed to reveal the mechanism(s) in different populations and in susceptible groups. This research might then lead to the best strategy how humans can protect their skin against the detrimental effects of ambient PM.

References

1. World Health Organization. Air quality guidelines for particulate matter, ozone, nitrogen dioxide and sulfur dioxide. Global update 2005. WHO/SDE/PHE/OEH/06.02. 2005. http://whqlibdoc.who.int/hq/2006/WHO_SDE_PHE_OEH_06.02_eng.pdf
2. Directive 2008/50/EC of the European Parliament and of the Council of 21 May 2008 on ambient air quality and cleaner air for Europe. 2008. http://eur-lex.europa.eu/LexUriServ/LexUriServ.do?uri=OJ:L:2008:152:0001:0044:EN:PDF
3. National Ambient Air Quality Standards (NAAQS). Washington, DC: US Environmental Protection Agency. www.epa.gov/air/criteria.html
4. Van Donkelaar A, Martin R, Brauer M, Kahn R, Levy R, Verduzco C, et al. Global estimates of ambient fine particulate matter concentrations from satellite based aerosol optical depth: development and application. Environ Health Perspect. 2010;118:847–55.
5. Katsouyanni K, Touloumi G, Samoli E, Gryparis A, Le Tertre A, Monopolis Y, et al. Confounding and effect modification in the short-term effects of ambient particles on total mortality: results from 29 European cities within the APHEA2 Project. Epidemiology. 2001;12:521–31.
6. Samet JM, Zeger SL, Dominici F, Curriero F, Coursac I, Dockery DW, Schwartz J, et al. The national morbidity, mortality and air pollution study part II: morbidity and mortality from air pollution in the United States. In: Research report from Health Effects Institute. 2002. www.cabq.gov/airquality/documents/pdf/samet2.pdf
7. Xu MM, Jia YP, Li GX, Liu LQ, Mo YZ, Jin XB, et al. Relationship between ambient fine particles and ventricular repolarization changes and heart rate variability of elderly people with heart disease in Beijing, China. Biomed Environ Sci. 2013;26:629–37.
8. Guo Y, Jia Y, Pan X, Liu H, Wichmann HE. The association between fine particulate air pollution and hospital emergency room visits for cardiovascular diseases in Beijing, China. Sci Total Environ. 2009;407:4826–30.
9. Madaniyazi L, Guo Y, Ye X, Kim D, Zhang Y, Pan X. Effects of airborne metals on lung function in inner Mongolian schoolchildren. J Occup Environ Med. 2013;55:80–6.

10. Li P, Xin J, Wang Y, Wang S, Li G, Pan X, et al. The acute effects of fine particles on respiratory mortality and morbidity in Beijing, 2004–2009. Int Environ Sci Pollut Res. 2013;20:6433–44.

11. Ranft U, Schikowski T, Sugiri D, Krutmann J, Krämer U. Long-term exposure to traffic-related particulate matter impairs cognitive function in the elderly. Environ Res. 2009;109(8):1004–11.

12. Schikowski T, Vossoughi M, Vierkötter A, Schulte T, Teichert T, Sugiri D, Fehsel K, Tzivian L, Bae IS, Ranft U, Hoffmann B, Probst-Hensch N, Herder C, Krämer U, Luckhaus C. Association of air pollution with cognitive functions and its modification by APOE gene variants in elderly women. Environ Res. 2015;142:10–6.

13. Larrieu S, Lefranc A, Gault G, Chatignoux E, Couvy F, Jouves B, et al. Are the short-term effects of air pollution restricted to cardiorespiratory diseases. Am J Epidemiol. 2009;169:1201–8.

14. Krämer U, Sugiri D, Ranft U, Krutmann J, von Berg A, Berdel D, et al. Eczema, respiratory allergies, and traffic-related air pollution in birth cohorts from small-town areas. J Dermatol Sci. 2009;56:99–105.

15. Morgenstern V, Zutavern A, Cyrys J, Brockow I, Koletzko S, Krämer U, et al. Atopic diseases, allergic sensitization, and exposure to traffic-related air pollution in children. Am J Respir Crit Care Med. 2008;177:1331–7.

16. Kim J, Kim EH, Oh I, Jung K, Han Y, Cheong HK, et al. Symptoms of atopic dermatitis are influenced by outdoor pollution. J Allergy Clin Immunol. 2013;132:495–7.

17. Vierkötter A, Schikowski T, Ranft U, Sugiri D, Matsui M, Krämer U, Krutmann J. Airborne particle exposure and extrinsic skin aging. J Invest Dermatol. 2010;130(12):2719–26.

18. Lefebvre MA, Pham DM, Boussouira B, Bernard D, Camus C, Nguyen QL. Evaluation of the impact of urban pollution on the quality of skin: a multicentre study in Mexico. Int J Cosmet Sci. 2015;37(3):329–38.

19. Lademann J, Richter H, Schanzer S, Knorr F, Meinke M, Sterry W, Patzelt A. Penetration and storage of particles in human skin: perspectives and safety aspects. Eur J Pharm Biopharm. 2011;77(3):465–8.

20. Bolzinger MA, Briançon S, Chevalier Y. Nanoparticles through the skin: managing conflicting results of inorganic and organic particles in cosmetics and pharmaceutics. Wiley Interdiscip Rev Nanomed Nanobiotechnol. 2011;3(5):463–78.

21. Donaldson K, Mills N, MacNee W, Robinson S, Newby D. Role of inflammation in cardiopulmonary health effects of PM. Toxicol Appl Pharmacol. 2005;207:483–8.

22. Fritsche E, Schäfer C, Calles C, Bernsmann T, Bernshausen T, Wurm M, et al. Lightening up the UV response by identification of the arylhydrocarbon receptor as a cytoplasmatic target for ultraviolet B radiation. Proc Natl Acad Sci U S A. 2007;104:8851–6.

23. Jux B, Kadow S, Luecke S, Rannug A, Krutmann J, Esser C. The aryl hydrocarbon receptor mediates UVB radiation-induced skin tanning. J Invest Dermatol. 2011;131:203–10.

24. Haarmann-Stemmann T, Esser C, Krutmann J. The Janus-Faced role of aryl hydrocarbon receptor signaling in the skin: consequences for prevention and treatment of skin disorders. J Invest Dermatol. 2015;135(11):2572–6.

25. Choi H, Shin DW, Kim W, Doh SJ, Lee SH, Noh M. Asian dust storm particles induce a broad toxicological transcriptional program in human epidermal keratinocytes. Toxicol Lett. 2011;200(1–2):92–9.

26. Tigges J, Haarmann-Stemmann T, Vogel CF, Grindel A, Hübenthal U, Brenden H, Grether-Beck S, Vielhaber G, Johncock W, Krutmann J, Fritsche E. The new aryl hydrocarbon receptor antagonist E/Z-2-benzylidene-5,6-dimethoxy-3,3-dimethylindan-1-one protects against UVB-induced signal transduction. J Invest Dermatol. 2014;134(2):556–9.

27. Brook RD, Rajagopalan S, Pope CA 3rd, Brook JR, Bhatnagar A, Diez-Roux AV, Holguin F, Hong Y, Luepker RV, Mittleman MA, Peters A, Siscovick D, Smith SC Jr, Whitsel L, Kaufman JD. American Heart Association Council on epidemiology and prevention, Council on the kidney in cardiovascular disease, and Council on nutrition, physical activity and metabolism. Particulate matter air pollution and cardiovascular disease: an update to the scientific statement from the American Heart Association. Circulation. 2010;121(21):2331–78.

POPs and Skin

10

M.M. Leijs, Janna G. Koppe, T. Kraus, J.M. Baron,
and H.F. Merk

Persistent organic pollutants (POPs) is the name of a group of ubiquitous man-made toxic environmental pollutants which have adverse effects on the environment and human health. The skin is a major target organ of toxic effects of POPs as well as signalling organ for the toxicity of these compounds. Chloracne is the most obvious signs of systemic contact with higher levels of POPs with chloracnegenic potential. First we will discuss chemical and toxicological properties of POPs and finally the pathophysiology of cutaneous diseases derived from POPs including acne, cancerogenesis, porphyria and others.

10.1 Persistent Organic Pollutants (POPs)

Persistent organic pollutant (POP) is the name of a group of ubiquitous man-made toxic environmental pollutants which have adverse effects on the environment and human health.

Twelve compounds of the POPs have been nominated by the Environmental Protection Agency (EPA) as being the most harmful [1]. They are called the dirty dozen (see Table 10.1): Most of the listed compounds, except for the dioxins (PCDD/Fs), are intentionally produced. Among them are

Table 10.1 The dirty dozen

Aldrin	Insecticide
Chlordane	Insecticide
DDT	Insecticide
Dieldrin	Insecticide
Endrin	Insecticide
Heptachlor	Insecticide
Hexachlorobenzene	Fungicide
Mirex	Insecticide
Toxaphene	Insecticide
PCBs	Used for their fire resistance, low electrical conductivity in electrical equipment
Dioxins (PCDDs) Dioxins (furans) (PCDFs)	Unintentionally formed during chemical, thermal, photochemical and enzymatic reactions

This table represents the original 12 POPs according to the Stockholm Convention

M.M. Leijs (✉) • J.M. Baron • H.F. Merk
Department of Dermatology and Allergology,
RWTH Aachen University, Pauwelstrasse 30,
Aachen, 52075 Germany
e-mail: mleijs@ukaachen.de

J.G. Koppe
Department of Paediatrics and Pulmonology,
Emma Children's Hospital, Academic Medical
Center, Meibergdreef 9, Amsterdam, 1105 AZ, The
Netherlands

Ecobaby Foundation, Hollandstraat 6,
Loenersloot 3634 AT, The Netherlands

T. Kraus
Institute of Occupational Medicine,
RWTH Aachen University, Pauwelstrasse 30, Aachen
52074, Germany

© Springer International Publishing Switzerland 2018
J. Krutmann, H.F. Merk (eds.), *Environment and Skin*,
https://doi.org/10.1007/978-3-319-43102-4_10

insecticides and fungicides used in agriculture and so entering the food chain. Although most of these insecticides are forbidden, it has been reported that they are still used in developing countries.

As a result of their persistency nowadays, the majority of the world population has measurable background levels in blood serum.

Newly added POPs include the flame-retardants polybrominated biphenyls (PBBs) especially hexabromobiphenyl (HBB), the herbicide chlorde-cone and the insecticide lindane also known as gamma-hexachlorocyclohexane (γ-HCH) and its sides products (α-hexachlorocyclohexane (α-HCH) and β-hexachlorocyclohexane (β-HCH)). Other important newly added POPs are some of the congeners of the polybrominated diphenyl ethers (PBDEs, a flame retardant) and pentachloroben-zene (PeCB, pesticide and flame retardant), as well as perfluorooctanesulfonic acid (PFOS) and per-fluorooctanesulfonyl fluoride (PFOSF) used as sur-factants in various industry and consumer products (electrics, hydraulic fluids and textiles), endosul-fans (insecticides used in wood preservatives) and hexabromocyclododecane (HBCD, a flame retardant).

Because of their persistency, these compounds bioaccumulate through the food web. Three major conventions have been grounded in order to reduce and gradually eliminate these persistent endocrine disrupting and carcinogenic compounds. One of them, the Stockholm Convention on Persistent Organic Pollutants, was adopted in 2001. The Convention built on the precautionary principle, principle widely used in environmental law.

10.1.1 PCBs and Dioxins: The Classical Causative Agents for Environmental Skin Diseases

Polychlorinated dibenzo-p-dioxins (PCDDs) and polychlorinated dibenzofurans (PCDFs) belong to the group of the dioxins. A PCDD molecule (see Fig. 10.1) exists of two benzene rings connected by two oxygen atoms. In PCDFs (see Fig. 10.2) the two benzene rings are connected by one oxygen atom. The carbon atoms may be attached to a chlorine atom in position 1–4 and 6–9. There are 75 possible PCDDs and 135 possible PCDFs. After

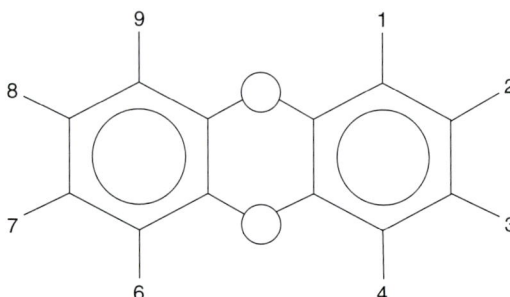

Fig. 10.1 Polychlorinated dibenzo-p-dioxins (PCDDs)

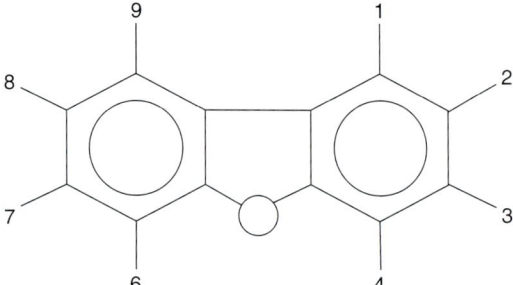

Fig. 10.2 Polychlorinated dibenzofurans (PCDFs)

several chemical accidents resulting in serious health effects, severe birth abnormalities and mortality, dioxins and PCBs are well known in the general public. Dioxins exert a certain kind of toxicity which can be measured in toxic equivalency factors (TEFs); the position of the chlorine atoms on the molecule determines the toxicity of each congener. PCDD/Fs containing one to three chlorine atoms are thought to have no significant toxicological significance. The most toxic dioxin is 2,3,7,8-TCDD; this compound is the reference for the development of the toxic equivalency factors (TEFs).

PCDDs and PCDFs can be formed during chemical, thermal, photochemical and enzymatic reactions. Several thermal processes leading to the formation of PCDDs and PCDFs have been identified. Sources of emission of PCDD/Fs are [2–4]:

– The incineration of municipal solid waste, sewage sludge, coal, peat, wood, hospital waste and hazardous waste
– Cigarette smoking
– Bleaching of wood paper and pulp
– Cement kilns
– Production of metals like steel and copper

- Contaminated commercial products such as chlorinated phenols and PCBs
- Wire reclamation
- Fossil fuel combustion, diesel heavy-duty trucks, cigarette smoke
- Backyard barrel burning
- Accidents in PCB-filled electrical equipment

10.1.2 The Main Source of These Compounds in Non-industrial Countries Was the Incineration of Waste [2]

Once released in the environment, the fate and behaviour of individual congeners depend on these physico-chemical properties, and higher chlorinated compounds are more lipophilic. As a result of this lipophilic character, these compounds are sensitive to absorb to organic compounds, and only small amounts dissolve in water. As a result of their persistent character in our environment, they bioaccumulate and exert a persistent influence on physiological processes in our body.

Polychlorinated biphenyls (PCBs) is the name of a group of 209 compounds (congeners) with a similar structure (see Fig. 10.3). In PCBs the position of the chlorines on the PCB molecule can be divided in the ortho-, meta- and para-position. The non-ortho-substituted congeners and some mono-ortho congeners can have a planar structure and are therefore sometimes named dioxin-like (dl) PCBs and can exert a certain kind of toxicity through the Ah receptor.

PCBs are mainly used as dielectric isolators in electrical equipment because of their fire resistance, low electrical conductivity, high thermal conductivity and high resistance to thermal degradation. From the 1930s the production of PCBs drastically increased worldwide, reaching a maximum in the 1970–1980s [5]. Since the 1930s large amounts of dioxins and PCBs have been released into the environment. Organisms, and ultimately humans, are exposed via ingestion (food, drinking water), via inhalation, and via dermal contact.

Ingestion is the main route (90%) of exposure to dioxins and PCBs in Europe, primarily through meat and meat products (23–27%), dairy products (17–27%) and fish (16–26%). However, dermal absorption and inhalation play also an important role in occupational exposure [6].

Due to the accumulating properties of these compounds, each step higher in the food chain increases the concentration of dioxins in an organism (bioaccumulation). Once ingested, dioxins and PCBs are primarily stored in adipose tissue and liver, the result of their hydrophobic nature. The first measurement of dioxins in humans was carried out in 1956, when a chemist contaminated himself with tetrachlorodibenzo-*p*-dioxin (TCDD) and tetrabromodibenzo-*p*-dioxin [7]. Currently, background levels of these compounds have been found throughout the industrialised world. Since the 1990s background levels of dioxins are decreasing due to better legislation and filters in waste incinerators [8].

How long they remain in the human body is measured using half-life, which is defined as the time it takes before the amount of a material is decreased by 50%. The half-life of each PCB and dioxin congener varies. The mean half-life of dioxins and PCBs in the human body is assumed to be 7–9 years [9], but may be shorter [10]. For PCBs this strongly depends on the amount of chloratoms on the molecule [11].

PCDDs/PCDFs and planar (dioxin-like) PCBs are usually discussed together because of their similar toxicity. They have the ability to bind the aromatic hydrocarbon receptor (AhR). The AhR is a multiprotein complex and a cytosolic transcription factor, which is in normal condition inactive and binds to other proteins called co-chaperones. It is present in a wide range of cell types and species. Bonding of these receptors is known to induce expression of cytochrome P45

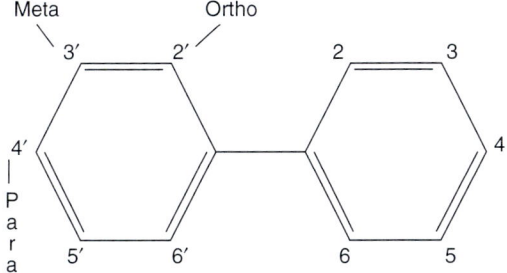

Fig. 10.3 Polychlorinated biphenyl (PCB)

by induction of the CYP1 gene family. In addition other xenobiotic metabolizing enzymes, transporter proteins and others of the AhR-gene battery are also induced [12, 13].

The Ah receptor can bond to ligands like dioxin-like compounds such as benzopyrene; however, there are also naturally occurring ligands like derivatives of tryptophan and bilirubin; however, these have a less affinity for the receptor than TCDD and rapidly degradates [14, 15].

Dioxin-like compounds (PCDDs/Fs and planar PCBs) are able to bind to the Ah receptor and are thereby considered to behave like anti-oestrogens. However, one study showed that some dioxin-like compounds can have oestro-genic properties as well [16]. To predict the health effects of PCDD, PCDFs and dioxin-like PCBs, a toxic equivalency score (TEQ) was developed, considering TCDD the most toxic compound having a toxic equivalency factor (TEF) as value 1 [17, 18].

The dose, which causes 50% mortality, is called the (lethal dose) LD50. This dose is different for several species. For example, LD50 values vary up to 5000-fold between guinea pig and hamster. For example, for TCDD in fish it is species dependent and ranges from 3000 to 16,000 pg/TCDD/g fish [19].

Although LD50 values vary up to 5000-fold between guinea pig and hamster, TCDD-induced aryl hydrocarbon hydroxylase (AHH) and ethoxyresurufin-O-deethylase (EROD) enzyme induction and reproductive toxicity is very similar in these two species [20].

Lipophilic chemicals like PCBs and dioxins can pass the placenta and via the vena umbilicalis go directly via the foramen ovale to the left side of the heart and in the aorta and to organs like the brain, pancreas and kidney. In addition, they are excreted in breast milk and thereby cause significant exposure to nursing offspring [21, 22]. Exposure during the foetal and nursing period of a child is considered to be the most sensitive exposure windows in terms of reproductive effects [23–26]. Adolescents, undergoing hormonal changes during puberty, are probably also at greater risk of susceptibility and therefore at higher risk with regard to environmental exposure health effects [27].

The non-dioxin-like congeners of PCBs may exert different effects. Phenobarbital-like oestrogenic and neurotoxic effects have been described [28, 29]. However the in vitro oestrogen receptor (ER) binding appears weak in some studies [30]. PCBs and their hydroxylated metabolites are also known for their effects on the thyroid regulatory pathway, both reducing and increasing serum thyroid hormone (T4) levels [31, 32].

10.2 Observed (Cutaneous) Health Effects and Chemical Incidents

High-dose effects in animals have shown lethality following a period of the *wasting syndrome*, a severe weight loss obvious within a few days after exposure.

Various studies have looked at the effects of the dioxin/PCB exposure in humans; the most famous study cohorts were based on highly exposed individuals as a result of chemical accidents, the most important accidents being the Yusho, Seveso, Yucheng and Agent Orange incidents.

10.2.1 The Yusho Incident

In 1968 a mass poisoning occurred in Japan after a rice bran oil produced by Kanemi Company in Kyushu started to become contaminated with PCBs and PCDFs. This oil was sold to poultry farmers for use as a feed supplement and also to consumers for use in cooking. After suppletion to the poultry, farmers started reporting that their poultry were dying due to apparent difficulty in breathing; altogether 400,000 birds died.

In the human who consumed this oil, a disease appeared, characterised by acne-like eruptions, pigmentation of the skin and eye discharge. The disease was named Yusho (oil disease). More than 1800 patients have been registered as having Yusho and around 300 deceased [33, 34].

10.2.2 Yucheng Disease

Similarly to Yusho, in 1978 a similar incident happened in Taiwan. A PCB mixture (Kanechlor 400,

500) was used as the heat transfer medium in the process of deodorisation and decolorisation of rice oil by a rice oil company in central Taiwan. The PCBs together with their heat-degraded byproduct (PCDFs) leaked into the rice oil and intoxicated 2000 people on Taiwan Island. The exposed population developed chloracne, hyperpigmentation, peripheral neuropathy and other symptoms, which were later called Yucheng disease [35, 36].

10.2.3 Seveso

In 1976 in Seveso, in a small chemical manufacturing plant approximately 15 km (9 mi) north of Milan in Italy, an explosion took place. Large amounts of 2,3,7,8-TCDD, the most toxic dioxin, were released into the environment by the explosion. Within days a total of 3300 animals were found dead, mostly poultry and rabbits.

The explosion resulted in major health consequences for humans as well, including malformations, miscarriages, respiratory problems, cancer and death, for the exposed adults and their children [37, 38].

10.2.4 Agent Orange

Agent Orange was a powerful mixture of chemical defoliants used by US military forces during the Vietnam War to eliminate forest cover for North Vietnamese and Viet Cong troops. The US programme of defoliation (1962-1972), codenamed Operation Ranch Hand, sprayed more than 54 million litres of herbicide contaminated with TCDD over South Vietnam. Besides the effects seen in the inhabitants of the region, a significant dose-response relationship with, a history of benign fatty tumours, chloracne, skin rashes with blisters and photophobia were seen in the veterans [39].

10.3 Epigenetics and PCBs

In Slovakia (Eastern Europe), the Chemko factory dumped a lot of PCBs in the Laborec River, and this area of Michalovce is one of the most polluted areas with PCBs. The impact on the population

was studied, and growth restrictions of babies, boys more than girls, were found together with a poor neurodevelopment and reduced levels of thyroid hormone; hearing impairments, with harmful effects on the outer hair cells of the cochlea; and endocrine disruption of several thyroid and metabolic disorders and diabetes. In an existing mother-child birth cohort among children at the age of 46 months, gene expression and relevant disease pathway analysis was conducted and related to the current PCB levels (15 congeners). A few notable genes, such as BCL2, PON1 and ITGB1, were significantly altered in this study, and the related pathway analysis explained involvement in disease processes like cardiovascular disease and cancer. Interestingly the melanocyte development and pigmentation signalling are mentioned with the molecules SOX 10, PIK3R5 and BCL2 [40]. The SOX 10 gene is known to be related to the Waardenburg syndrome in which pigmentational disorders of the skin, hair and eyes are seen as well as hearing problems and uveal melanoma.

10.3.1 Flame Retardants: PBBs and PBDEs

Polychlorinated diphenyl ethers (PBDEs) are a type of brominated flame retardants (BFRs).

These compounds are commonly used as flame retardants in textiles, carpets, plastics and electrical equipment. PBDEs are seeded into but not covalently bound into polymer matrices. As a result of this, they diffuse out of the polymer matrix and become airborne and are widely dispersed [41, 42]. When a fire develops, bromine radicals in this molecule are released as a result of thermal energy. These radicals can decrease flame and reduce heat and carbon monoxide production [43]. They are forbidden in Europe, but not all congeners are forbidden in the United States. Although PBDEs are structural similar to PCBs and PCDD/Fs, no human or animal studies reported effects on the skin following PBDE exposure. In vitro study on human and rat skin showed that systemic (occupational) exposure to PBDEs through dermal absorption is very low in humans (3.1% and 1.9% for, respectively, tetrabromodiphenyl ether (TeBDE) and pentabromodiphenyl ether (PeBDE)) [44].

Although health effects of these compounds are not so well studied like dioxins and PCBs, possible endocrine disruption is reported [45] [46–49], as well as effects on the thyroid hormone [50–55], and neurotoxic effects [56–60] as well as effects on our immune system [61–63] and lung function [64].

Other brominated flame retardants are the polybrominated biphenyls. They are similar to PCBs but are a brominated analogue. PBBs are as well used as flame retardants. They are added to plastics used in products such as home electrical appliances, textiles and plastic foams. In 1973 an environmental accident occurred in northern Michigan in which large amounts of the toxic PBB (FireMaster BP-6) were accidentally mixed with livestock feed that was distributed to farms in Michigan, USA. It subsequently entered the human food chain of the entire state of Michigan.

Health effects, including cutaneous problems, were noted in contaminated animals and among exposed farmers some months after the contamination. Cutaneous abnormalities included: halogen acne, hair loss (alopecia), redness of the skin including irritant or allergic contact dermatitis, skin peeling and scaling, itching, increased sweating and nail dystrophy [65].

10.4 Cutaneous Symptoms of Exposure to POPs

10.4.1 Chloracne

A Summary of Cutaneous Symptoms Following Exposure to POPs Is Given in Table 10.2

Chloracne: Chloracne is an environmental skin disease characterised by an acne-like eruption of comedones (blackheads and whiteheads), cysts (varying from 1 to 10 mm) and pustules. It is the most obvious signs of systemic contact with higher levels of POPs with chloracnegenic potential.

Von Bettman (1897) [77] and Herxheimer [78] were on the first to describe this condition in German industrial workers. The pioneering contributions of KH Schulz and Kimmig proved that this disease is related to halogenated POPs in particular TCDD [79, 80]. This skin disease was named "chloracne" as a result of the clinical similarities with acne vulgaris and because they suggested this condition was caused by chlorine exposure [81–83].

This environmental skin disease is characterised by an acne-like eruption of comedones (blackheads and whiteheads), cysts (varying from 1 to 10 mm) and pustules, which is caused by an alteration of keratinisation of the pilosebaceous unit [84, 85].

Observation of highly exposed individuals showed that chloracne starts with an erythema (and oedema) of the face [86]. The typical comedones are formed after some days. In highly exposed individuals sometimes even all follicles on the face are turned into blackheads, the skin might show then a slate grey appearance [87].

It is solely found in highly exposed individuals and is therefore seen as the hallmark signalling of acute exposure to dioxins or other chloracnegens. This is proven in studies on higher exposed (PCBs, PCDD/PCDFs) individuals. Even 82–89% of the studied subjects reported dermatological manifestations [88]. In workers at trichlorophenol plants, where dioxins like TCDD are formed as byproducts, almost all workers had chloracne, and only about one third showed systemic intoxication [89].

Another example of chloracne is the poisoning of Victor Yushchenko. After a dinner in Kiev (2004), he became seriously ill, with symptoms of acute pancreatitis. His condition showed many unusual features, and 3 weeks later he developed a severe disfiguring rash. Investigations showed that he was poisoned by TCDD [90]. The levels in his blood serum (108,000 pg/g lipid weight) were more than 50,000-fold greater than those in the general population [75].

In another study two young women developed acneform lesions after moving into an office building. Clinically, the most affected woman was suspected of having acne fulminans, an acute febrile ulcerative acne conglobata; however, the acne did not response on high dose of oral corticosteroids, which acne fulminans normally does [71].

How do we distinguish chloracne from the classic acne (acne vulgaris)?

Clinical features: In chloracne there is not much inflammation visible. Lesions in acne

Table 10.2 Overview of human studies on different chloracnegens and found skin abnormalities

Study	Name of POP	Cutaneous symptoms
Ouw et al. 1976 [66] 34 workers of capacitor factory	PCBs (Aroclor 1242)	Burning sensation (face + hands), chloracne, eczematous rash, persistent body odor (unplesant smell)
Fischbein et al. 1979 [67, [68]] 326 workers of 2 U.S. capacitor factories	PCBs (Aroclor 1016, 1242, 1221, 1254)	Skin rash, burning sensation, acne, dryness and thickening of the skin, hyperpigmentation oculodermatological signs (oedema of upper eyelid, injected conjunctiva, eye discharge, hyperpigmentation, enlargement of Meibomian glands)
Chanda et al. [65] (Michigan accident, 996 exposed individuals, 149 control)	PBB	Halogen acne, folliculitis, hair loss (diffuse alopecia), skin redness (irritant or allergic contact dermatitis), skin peeling and scaling, itching, increased sweating and increased growth of fingernails and toenails
Cheng, Coenraads 1991 [72] 109 workers	Pentachlorophenol contaminated with dioxins (PCDD/Fs)	Chloracne, itchiness of the skin, hyperpigmentation, porphyria cutanea tarda
Piacitelli, L et al. 2001 [73] 3,538 workers	TCDD, 2,4,5-trichlorophenol and derivatives	Chloracne
Saurat et al. [69] Sorg et al. [75] One person contaminated with five million fold normal dose of TCDD	TCDD	Chloracne up to 40% of total body surface, hamartomas, involution of sebaceous glands
Geusau et al. [70, 71] 107 Two Women	TCDD	Severe chloracne, painful cyst (entire skin surface), palmoplantar keratoderma, granuloma annulare like lesions, distal onycholysis, brown/grey hyperpigmentations, hypertrichosis
Yusho incident Hsu et al. [33], Yoshimura [34], Reggiani and pruppacher 1985	PCBs/PCDFs	Acne-like eruptions, itchiness and sweating of the palms, pigmentation of the nails, skin, mucous membranes, distinctive hair follicles. Oculodermatological signs (hypersecretion of Meibomian glands, swelling of the eyelids, hyperpigmentation of the conjunctivae
Chen et al. [35], Rogan et al. [36] Yucheng	PCBs/PCDFs	Chloracne, hyperpigmentation
Bertazzi, Domenico [37] Seveso Caramaschi et al. [74] 164 exposed children (Seveso)	TCDD	Chloracne
Agent Orange Stellman et al. [39]	2,4,5-Trichlorophenoxyacetic acid, 2,4-dichlorophenoxyacetic acid, TCDD	Chloracne, skin rashes with blisters and photophobia
Leijs et al. [107] 304 workers from a transformer recycling company	PCBs	Hyperpigmentations, unspecific facial pa pules and pustules, chloracne (1 worker), mycosis fungoides (1 worker)
Michielsen et al. [76] (accidental poisoning of 4000 inhabitants in Turkey)	Hexachlorobenzene	Prominent bullous skin lesions, mainly in sun-exposed skin areas, that healed with severe mutilating scars (porphyria cutanea tarda)
Loomis et al. 138,905 workers of electrical power company	PCB/PCDF	Higher incidence of malignant melanoma
Gallagher et al. (115) 80 patients 310 controls	PCB	Higher risk (6-fold) of developing malignant melanoma with higher PCB levels

vulgaris are therefore much more reddish. In chloracne there are mainly comedones and cysts, and less pustules, while in acne vulgaris comedones are mixed with papules and pustules. Another important feature of chloracne is the decrease of sebum production, which comes with a dry skin. In patients with acne vulgaris, the sebum production is increased, which rather gives features of a fatty skin.

Anamnesis: Acne vulgaris affects mostly adolescents and young adults, while chloracne appears in every age group. A clear view should be given about to which chemicals an individual is exposed and in which time span the lesions occurred. Chloracne lesions are known to improve after elimination of the chloracnegen. The time until improvement depends on the half-life, chloracnegen potency and dosis of the chloracnegen and can be from weeks till decades.

Localisation: In chloracne the retroauricular and malar areas of the face are mainly involved, while the nose is mostly spared; also axillae, groin and extremities are frequently involved, while in acne vulgaris the nose is not speared (T-zone) and lesions are found on the upper back and chest.

Microbiological and histological features: Chloracne is normally sterile; only after secondary infection after manipulation bacteria can be found [91]. In acne vulgaris, however, *Propionibacterium* acnes and *Propionibacterium* granulosum can be cultured. Histologically, the epithelial lining of chloracne comedones is thicker (acanthotic) than in acne vulgaris. Chloracne shows more often a complete disappearance of the sebaceous glands. In acne vulgaris they mostly remain intact or are hypertrophied.

Not every POP is a chloracnegen, and whether an individual will develop chloracne depends on several factors:

1. Dosage of exposed chloracnegens
2. Chloracnegenic potency of the compound
3. Individual susceptibility of the exposed subject

Examples of chloracnegens are the already mentioned PCBs and dioxins (PCDD/Fs), and other well-known chloracnegens are polychloro-azobenzene and polybrominated biphenyls (PBBs). The contribution of chlorinated phenols and chloronaphthalenes is unclear, since they are mostly contaminated with PCBs or dioxins. Although these compounds can exert toxicological effects through the aromatic hydrocarbon receptor (AhR), it is not clear if they really contribute to the chloracne.

In order to achieve an effect through the AhR, the molecule should have two benzene rings and halogenated molecules in a lateral position to give a planar structure. However, the exact mechanism responsible for chloracne is not well understood, since not all compounds which exert effects on the AhR give chloracne.

Considering the individual susceptibility of exposed subjects, not much information is available. Interestingly a study on the exposed individuals of the Seveso incident showed that persons with lighter hair colour and younger of age were more susceptible to developing chloracne [92].

Some studies suggest an effect on the vitamin A metabolism in the skin could be an explanation for developing chloracne. Many studies reported a depletion of hepatic vitamin A storage and alteration in plasma retinol levels. Explanation for this could be enhanced breakdown by P450 isozymes or by UDP-glucuronyltransferases. As another explanation, disturbances of the formation of the retinol-binding protein have been suggested [93].

Most probably the pathogenesis of chloracne is caused by multiple reactions that are related to cell proliferation. One study on 12 biopsies of chloracne patients showed an induction of p-EGFR(epidermal growth factor receptor), p-MAPK (mitogen-activated protein kinase) and CK17(cytokeratin-17) mRNA and protein, which was not found in 12 controls. Chloracnegens bind the AhR which activates tyrosine kinase (PTK) activity and then enhances the RAS protein and MAPK phosphorylation enzyme cascade, which alter the expression of growth factors like the epidermal growth factor (EGF). Keratinocyte transglutaminase (TGk) mRNA and protein, a member of the transglutaminase family, were detected in both groups, but showed a different distribution.

In chloracne tissues positive signals were found mainly in the stratum granulosum and stratum spinosum and for controls mainly in the stratum granulosum. Mutation of this gene results in lamellar ichthyosis. In this study it was suggested that in the human skin, the activation of MAPK pathway and upregulation of CK17 and TGK may play roles in the pathogenesis of chloracne related to dioxin exposures [94].

Another pathway which might be important in the pathophysiology of chloracne is the transcription factor Nrf2, a key regulator of cellular stress response (ROS detoxification). Effects of Nrf2 on the skin however are controversial, on one side UvB-induced ROS damage and keratinocyte apoptosis were reduced in transgenic mice expressing caNrf2 in keratinocytes, and on the other side in vivo these mice induced infundibular acanthosis, hyperkeratosis and cyst formation. These features were linked to upregulation of epigen (a growth factor and novel Nrf2 target) as well as secretory leukocyte peptidase inhibitor (Slpi) and small proline-rich protein 2d (Sprr2d). SLPI, SPRR2 and epigen were also upregulated in dioxin-stimulated keratinocytes. More worrying is a correlation between Nrf2 with increased malignancy and chemoresistance of tumour cells and Nrf2 activation mutations in cutaneous squamous cell carcinomas [95].

Studies showed that effects induced by TCDD differ in different skin structures. Epidermis and infundibulum undergo hyperplasia, sebaceous and sweat glands lose their secretory activity and are replaced by keratinizing cells, while the lower portion of the follicle involutes. Because of the persistency of chloracne, it was suggested that involvement of the stem cells takes place, which also would explain the delayed onset of chloracne [82].

In vivo several major targets of TCDD were reported: cytokines (interleukin-1*B*, TNF (tumour necrosis factor)), growth factors (EGRF, TGF-*B* (transforming growth factor) and different genes in the apoptotic (Fas ligand, caspases, genes of the Bcl-2 family)) and angiogenic pathways (vascular endothelial growth factor (VEGF) and the plasminogen activator cascade) and angiogenin, which also influences proliferation and differentiation pathways in the skin [82].

10.4.2 Histological Features of Chloracne

Although a biopsy is helpful in some cases, the histopathological changes in long-standing cases are not specific for chloracne [96]. TCDD-treated epidermal equivalents have been shown to develop key features found in chloracne histology including: hyperkeratinisation of the stratum corneum and a thinner viable cell layer [97]. Other histological abnormalities seen in chloracne are follicular infundibular dilatation, comedones and follicular horny cysts. Acute chloracne exhibits squamous metaplasia of the sebaceous glands and lamellae of keratin-like material in its ducts [93].

10.5 Therapy of Chloracne

Chloracne is documented as being highly therapy resistant. The poor understanding of its pathogenesis and molecular downstream pathways limits a therapeutical approach.

Regular topical therapy as used in acne vulgaris is documented not to work effectively. Some studies even say that the only therapy for chloracne is eliminating the source of chloracnegen [82, 98]. Several studies confirmed that moderate and severe forms of chloracne are not sensitive to vitamin A therapy [82]. In contrary some studies say that topical retinoids (comedolytic agents) are effective in the treatment of chloracne [99]. Systemic isotretinoin was not effective in one highly exposed female patient [70].

Orally administrated corticosteroids have also proven to be ineffective in the treatment of chloracne [70, 71], and it was even shown to aggravate the condition in early stages [91].

However, dermabrasion, incision and drainage of cyst as well as extraction of comedones and light electrodesiccation might have positive effects [100].

In the study on the poisoning of Victor Yushchenko, non-steroidal anti-inflammatory drugs and systemic steroids were not effective to improve his condition with severe chloracne. The patient received thereafter three infusions of infliximab and was then switched to adalimumab.

Although generation of new hamartomatous lesions abruptly declined from month 28, the extent to which this improvement was driven by TNF-α blockade versus the repeated interventional treatment (incision of hamartomas, dermabrasia mechanical dermabrasion and multiple micro punch extraction/aspiration techniques) is unclear [69].

Studies have been performed to eliminate chloracnegens from the blood and fatty tissues, which can be archived by the synthetic fat substitute Olestra [101]. The sucrose backbone of Olestra which is able to bind eight fatty acids is able to accelerate the faecal secretion of chloracnegens since there is no gastroenteral absorption and chloracnegens are highly soluble in it. However, possible side effects of Olestra have to be considered. One mice study showed even a 30-fold increase in the rate of excretion of hexachlorobenzene after suppletion of Olestra combined with caloric restriction [102]. In a study on two highly (TCDD) exposed women with chloracne, the excretion rate (half-life) was enhanced by using Olestra containing potato chips from 7–9 to 1.5–2.9 years. In this study LDL apheresis was performed twice a week in order to reduce the TCDD body burden, corresponding to the elimination of blood fat. However, the amount of excreted TCDD by using this method was not high enough and comparable to faecal excretion [103].

10.6 Hyperpigmentations

The formation of hyperpigmentations is after chloracne the main cutaneous sign following exposure to some of the mentioned chloracnegen POPs [104–106]. Hyperpigmentations are documented, not only after dioxin exposure but also after PCB exposure [107]. These hyperpigmentations are not only located on the skin but also on the gingiva after PCB and dioxin poisoning [108]. Also in the Yusho study (PCB and PCDF exposure), besides acneform eruptions, pigmentation of the skin and nails was found [33, 34].

Another example is the Yu-cheng study, after the poisening, 39 babies from highly PCB and PCDF exposed mothers showed severe hyperpigmentations, and 8 of them died of pneumonia, bronchitis, sepsis and premature and congenital weakness [35, 36].

In workers of a pentachlorophenol (PCP) factory, a wood-preserving agent contaminated with TCDD, almost all workers had hyperpigmentations in the face and reported an itchy skin [93].

Above-mentioned studies imply that formation of hyperpigmentation is partly due to a direct effect on the skin (most probably as part of a pigmented contact dermatitis) as well as systemic effects. In vitro exposure of normal human melanocytes to 2,3,7,8-tetrachlorodibenzo-p-dioxin (TCDD, the most toxic dioxin) results in activation of the AhR signalling pathway and an AhR-dependent induction of tyrosinase activity, the most important enzyme of the melanogenic pathway. The total melanin content was elevated in TCDD-exposed melanocytes [109].

In zebrafish, exposure to TCDD induced slower regeneration of amputated fins with hyperpigmentation on the newly formed fins. The formation of hyperpigmentation was reproducible in the experiment. The AhR pathway was identified as an important pathway in the regeneration in this study [110].

Cytochrome P4501A1 induction by TCDD has been associated with oxidative stress and DNA damage in mammals [111, 112]. TCDD exposure results in enhanced terminal differentiation and a decrease in epidermal growth factor (EGF) binding in normal human epidermal cells [113]. The differentiation pattern of human keratinocytes is also altered by TCDD exposure [97], suggesting that the mechanism of hyperpigmentation is tightly tied to altered differentiation [111, 112].

10.7 Porphyria

One other important visible skin symptoms that are reported after exposure to chloracnegens is porphyria cutanea tarda, a skin disease which results from a deficiency of the enzyme uroporphyrinogen decarboxylase (UROD), and is marked by the formation of erosions, blisters and higher skin sensitivity [93]. Hexachlorobenzene or HCB is another POP, known for effects on the skin. It is a highly persistent environmental chemical that has been used in the past as a fungicide. Nowadays emission into the environment occurs as a waste byproduct of chemical processes. An accidental poisoning of

4000 inhabitants in Turkey took place after consumption of HCB-treated seed grain in 1955–1959. HCB inhibits UROD resulting in sign and symptoms of porphyria cutanea tarda. Hepatic porphyria developed and patients showed prominent bullous skin lesions, mainly in sun-exposed skin areas, that healed with severe mutilating scars. The skin lesions are studied in rats and suggest a specific involvement of the immune system [76]. In addition this observation led to the development of a rat model of porphyria cutanea-like disease [114].

10.8 Other Dermatological Symptoms

10.8.1 Hyperkeratinisation and Other Dermatological Symptoms

In a woman who was intoxicated with TCDD besides the usual chloracneform lesions, acral granuloma annulare-like lesions and distal onycholysis (loosening of the nail from its nail bed) as well as brownish/grey hyperpigmentations were found. Later signs of hypertrichosis were reported. In addition, she developed punctate keratoderma-like lesions on the palms and soles. These lesions were negative for human papillomavirus and histologically characterised by cone-shaped hyperkeratoses invaginating, but not penetrating, into the dermis. Both clinically and histologically these alterations are essentially indistinguishable from what is described as keratosis punctata palmaris et plantaris (KPPP) [115].

10.8.2 Pruritus

Pruritus, or itchiness of the skin, has been reported in several studies after exposure to PCBs, dioxins and pentachlorophenol [93, 116].

In animal study, TCDD in combination with distilled water or acetone/olive oil application caused a significant increase in scratching behaviour. Furthermore, nerve growth factor (NGF) content in the skin increased significantly. Administration of histamine H1 receptor antagonist had no effect. After repeated administration for 7 days, the histamine H1 receptor antagonist olopatadine significantly inhibited scratching behaviour.

This study concluded that TCDD is not a pruritogen but causes alloknesis (itchy skin) [117].

10.8.3 Oculodermatological Findings

Oculodermatological signs and symptoms are as well as chloracne important diagnostic hallmarks for high internal levels of POPs like dioxins and PCBs. Hypersecretion of the Meibomian glands has been discribed, as well as swelling of the eyelids, and hyperpigmentation of the conjunctivae. These abnormalities were found in 60–85% of the affected individuals in the above-mentioned Yusho accident [118].

10.9 Malignant Melanomas and Non-pigmentated Cutaneous Neoplasia and Exposure to POPs

Some studies showed an elevated risk of developing malignant melanoma in workers suspected to be exposed to organohalogen compounds. In one study in the United States (1975–1980), the melanoma incidence in habitants (10–49 years) in a southwest Georgia town was significantly increased compared with US rates (expected = 9; observed = 41) and Atlanta rates (expected = 13; observed = 41). When 36 malignant melanoma patients were compared with 74 controls, the probability of developing malignant melanoma was related to a history of melanoma in family members ($P = 0.063$), skin sensitivity to sun exposure ($P = 0.016$), pre-existing pigmented nevi ($P = 0.005$) and exposure to sick animals ($P = 0.055$) and to pesticides in nonoccupational settings ($P = 0.059$) [119].

In one other study among 138,905 men employed >6 months (1950–1986) at five electrical power companies in the United States, the mortality from malignant melanoma of the skin was increased among men with the longest employment in jobs potentially exposed to PCB insulating fluids [120].

A study on the Agent Orange veterans showed a higher incidence of malignant melanoma and other types of cancer among the veterans compared to the normal population. In addition the veterans had a tendency to have lesions in exposed areas. Most veterans experienced pruritus. In this study, mycosis fungoides (MF) patients with a history of Agent Orange exposure differed significantly from those without exposure to Agent Orange in demographic and clinical characteristics; in addition there was a higher frequency (33.3%) of mycosis fungoides palmaris et plantaris [121, 122].

In one Canadian study, the relationship between several organohalogen compounds (14 PCB congeners and 11 pesticides) and the risk of cutaneous malignant melanoma was calculated in 80 patients with malignant melanoma and 310 controls. After correction for confounding factors, there was an association between PCB levels and the risk of developing a cutaneous malignant melanoma (a sixfold increase). The strongest relation was surprisingly with ndl-PCBs (sevenfold increase) compared to 2.8 for dl-PCBs. In this study, there was also a weak nonsignificant increase with p,p'-DDE. It is suggested that PCBs act as a tumour promotor, which might promote the transformation of melanocyte clones (nevi) [110, 123]. Another plausible explanation is the effect of dioxins on the immune cells, since a higher incidence of malignant melanoma has been reported in studies on immunocompromised patients.

10.10   Immunity, Dioxins and PCBs

The importance of the discovery of the role of cells in immunity became clear in the 1960s of the last century. It was Paul Langerhans, who for the first time the Langerhans cell described already in 1868, and finally Rudolph Baer and Ina Silberberg found that these cells are sentinels of the immune system in the skin [124].

In 1986 the term "skin-associated immune system" or cutaneous immune system is launched, and a skin-specific immunology was described. About half of the cells of the skin belong to the immune system, and many skin disorders are based on disruption of the homeostasis of this system. Examples of the immune-modulatory effects are corticosteroids and phototherapy with short-wave ultraviolet light (UVB). The immuno-

suppressive action is related to an inhibitory effect on the production of a cytokine. And also corticosteroids have inhibitory effects at the level of cytokines [125].

Immunity in humans can be divided in the innate immune system and the adaptive immune system. Between these two systems are clear differences, and the adaptive immunity response is highly specific to a micro-organism like a bacteria or virus or parasite and is known for its memory function. The innate immunity is not specific, is constantly controlling, can immediately attack and doesn't have a memory. Cells of the innate immunity like the polymorphonuclear cells (PMN or also named neutrophil) are necessary to activate the cells of the adaptive immunity. The adaptive immunity can be divided in a cellular response involving T-lymphocytes and a humoral response involving proteins in the blood and other body fluids. The cellular response are T-cells activated upon contact with an antigen, presented by, for instance, a Langerhans cell, and become activated cytotoxic T-cells killing micro-organisms. The humoral response is cooperation between T-cells and B-cells generating immunoglobulins directed to the invader. B-cells develop into antibody-secreting plasma cells with the help of the T-helper cells after antigen presentation by an antigen-presenting cell. The response of the adaptive immunity takes about 7–10 days to develop fully. In the first period of an infection, the cells of the innate immunity do the work.

The innate immune response is unspecific and forms the first line of defence. In early childhood especially in the interval between the decline and loss of the passively acquired antibodies from the mother and the acquisition of a mature adaptive immune system, the innate immunity needs to function optimally. The cells belonging to the innate immunity are the natural killer cells, monocytes and macrophages; PMN is also named neutrophils, basophils and eosinophils. Of the total number of white cells, the percentage of the PMNs is the highest. The cells of the innate immunity all originate from the myeloid progenitor cell, except the natural killer cells. All these cells act immediately with the help of proteins available in body fluids like the complement proteins to mark the invader and phagocyte the micro-organism and kill it with the reactive oxygen species (ROS) activity.

Blood platelets are also involved in the first line of defence and are often the first ones that are decreased in numbers during a beginning infection. Blood platelets also develop from the myeloid progenitor cell via the megakaryocytes. Parts of the pathways are known how the development takes place from the stem cell to the different forms of white cells, red cells and blood platelets, but not all.

PCBs and dioxins have negative immunosuppressive effects on the innate immunity, especially on the number of polymorphonuclear leukocytes (PMNs) and the ROS activity of these cells in early life and on the thrombocytes probably disturbing a smooth development of the immune system of the infant and young child.

Negative effects of environmental pollutants like dioxins and PCBs are most rigorous in early life. In the foetus the immunologic systems are present, both the innate one and the possibilities for an adaptive immunity enhanced by the passive transfer of immunoglobulins of the mother. Environmental pollutants are able to pass the placenta and so influence the normal foetal immunological development.

In Western Europe the background levels to dioxins and PCBs were relatively high in the years 1975–2000, and effects of the exposure prenatally via the placenta and postnatally by breastmilk on the immune system were studied. A significantly lower number of PMNs at 7 days of age in the prenatally higher to dioxins exposed group and a lower number of thrombocytes at 11 weeks in the babies with a higher exposure via breastmilk was found in a study performed in Amsterdam. In the adaptive immunity at the age of 8 years of the Amsterdam-Zaandam cohort and effects were found related to the lactational dioxin exposure: more $CD4^+$ (T-helper) cells as well as an increase in CD45RA cells (naïve T-cells).

Even at very low background levels, a suppression of the PMNs was seen, related to a rather low level of current dioxin-like PCBs (congener 77, 126, 169) of 2.2 TEQ pg/g fat in serum. The effect on the decreased number of PMNs during adolescence might be correlated to an increased bone marrow sensitivity to dioxin-like compounds and PCBs, resulting from the perinatal exposure, a phenomena described after perinatal exposure to arsenic and lead [126].

In addition, a persistent negative effect on the thrombocytes was seen in 7–12-year-old children in relation to the lactational dioxin exposure by breastmilk. This indicates an influence at the megakaryocyte and at the stem cell level, which is likely because from the stem cell and the myeloid progenitor cell the other above-mentioned "dioxin-sensitive" cells of the innate immunity are formed. An effect of dioxins on the number of thrombocytes was also found in Japanese workers with higher dioxin levels and in two Austrian women accidentally intoxicated with high concentrations of dioxins.

In a cross-sectional study in the VS part of the NHANES cohort of 2003–2004, also a decreased number in PMNs and white cells in general and in thrombocytes is found in relation to both the current dioxin-like and non-dioxin-like PCBs in serum [127].

In animal studies the finding of a persistent negative effect of perinatal dioxin exposure on the immune function is studied on the active T-cell population. Mice were exposed to an oral dose of TCDD that was so low that no hypocellularity in the thymus or bone marrow or other signs of toxicity were seen in the mother. In the adult offspring, durable changes in the responsive capacity and differentiation of $CD4^+$ T-cells are found [128]. The influence of dioxins impairing differentiation of normal human epidermal keratinocytes in a skin equivalent model is also demonstrated after a case study of poisoning of two secretaries [129].

PCBs and dioxins are immune-modulating chemicals with both immunosuppressive and enhancing effects. Especially the innate immunity is vulnerable. Effects on the immune system might play an important role in the pathogenesis of environmental toxicant-induced skin disease.

Conclusion

That some of the POPs can exert cutaneous effects is clear. Chloracne is not the only cutaneous symptom of high exposure to chloracnegens: also hyperpigmentations (skin, nails and gingiva), porphyria, pruritus, hypersecretion of the Meibomian glands and a higher change of developing malignant melanoma as well as lymphomas have been reported.

Although many studies are performed in the last 30 years in order to identify the pathophysiological mechanism of chloracne, the exact molecular mechanism is complex and remains unclear. In addition, studied cohort represents mostly human exposed to a mixture of chloracnegens and other POPs (pesticides, flame retardants), which makes it almost impossible to isolate the individual effects of these compounds in human beings.

A worrying effect of POPs on the skin is the higher incidence of malignant melanomas as well as the higher incidence of lymphomas, which is seen in (higher) exposed individuals. Here again a plausible explanation remains unclear. A negative influence of POPs on the immune system or on the melanogenesis might be an explanation.

References

1. Bailey M: U.S.Environmental Protection Agency. Persistent organic pollutants: a global issue, a global response. 2009. http://www2.epa.gov/international-cooperation/persistent-organic-pollutants-global-issue-global-response.
2. Olie K, Vermeulen PL, Hutzinger O. Chlorodibenzo-p-dioxins and chlorodibenzofurans are trace components of fly and flue gas of some municipal incinerators in The Netherlands. Chemosphere. 1977;6:455–9.
3. Rappe C. Analysis of polychlorinated dioxins and furans. Environ Sci Technol. 1984;18:78A.
4. World Health Organisation. PCDD and PCDF emissions from incinerators for municipal sewage sludge and solid waste – evaluation of human exposure. WHO Environmental Health Education. Copenhagen: WHO; 1987.
5. de Voogt P, Brinkman UAT. Production, properties and usage of polychlorinated biphenyls. In: Kimbrough RD, Jensen AA, editors. Halogenated biphenyls, terphenyls, naphtalenes, dibenzodioxins and related products. 2nd ed. Amsterdam: Elsevier; 1989. p. 3–45.
6. Baars AJ, Bakker MI, Baumann RA, Boon PE, Freijer JI, Hoogenboom LA, et al. Dioxins, dioxin-like PCBs and non-dioxin-like PCBs in foodstuffs: occurrence and dietary intake in The Netherlands. Toxicol Lett. 2004;151(1):51–61.
7. Baughman RW. Tetrachlorodibenzo-p-dioxins in the environment: high resolution mass spectrometry at picogram level. Cambridge: Harvard University; 1974.
8. Leijs MM, van Teunenbroek T, Olie K, Koppe JG, ten Tusscher GW, van Aalderen WM, et al. Assessment of current serum levels of PCDD/Fs, dl-PCBs and PBDEs in a Dutch cohort with known perinatal PCDD/F exposure. Chemosphere. 2008;73(2):176–81.
9. Pirkle JL, Wolfe WH, Patterson DG, Needham LL, Michalek JE, Miner JC, et al. Estimates of the half-life of 2,3,7,8-tetrachlorodibenzo-p-dioxin in Vietnam veterans of operation ranch hand. J Toxicol Environ Health. 1989;27(2):165–71.
10. Aylward LL, Brunet RC, Starr TB, Carrier G, Delzell E, Cheng H, et al. Exposure reconstruction for the TCDD-exposed NIOSH cohort using a concentration- and age-dependent model of elimination. Risk Anal. 2005;25(4):945–56.
11. Seegal RF, Fitzgerald EF, Hills EA, Wolff MS, Haase RF, Todd AC, et al. Estimating the half-lives of PCB congeners in former capacitor workers measured over a 28-year interval. J Expo Sci Environ Epidemiol. 2011;21(3):234–46.
12. Denison MS, Nagy SR. Activation of the aryl hydrocarbon receptor by structurally diverse exogenous and endogenous chemicals. Annu Rev Pharmacol Toxicol. 2003;43:309–34.
13. Nebert DW, Karp CL. Endogenous functions of the aryl hydrocarbon receptor (AHR): intersection of cytochrome P450 1 (CYP1)-metabolized eicosanoids and AHR biology. J Biol Chem. 2008;283(52):36061–5.
14. Adachi J, Mori Y, Matsui S, Takigami H, Fujino J, Kitagawa H, et al. Indirubin and indigo are potent aryl hydrocarbon receptor ligands present in human urine. J Biol Chem. 2001;276(34):31475–8.
15. Sinal CJ, Bend JR. Aryl hydrocarbon receptor-dependent induction of cyp1a1 by bilirubin in mouse hepatoma hepa 1c1c7 cells. Mol Pharmacol. 1997;52(4):590–9.
16. Ohtake F, Takeyama K, Matsumoto T, Kitagawa H, Yamamoto Y, Nohara K, et al. Modulation of oestrogen receptor signalling by association with the activated dioxin receptor. Nature. 2003;423(6939):545–50.
17. Safe S. Polychlorinated biphenyls (PCBs), dibenzo-p-dioxins (PCDDs), dibenzofurans (PCDFs), and related compounds: environmental and mechanistic considerations which support the development of toxic equivalency factors (TEFs). Crit Rev Toxicol. 1990;21(1):51–88.
18. van den Berg M, Birnbaum LS, Denison M, De VM, Farland W, Feeley M, et al. The 2005 World Health Organization reevaluation of human and mammalian toxic equivalency factors for dioxins and dioxin-like compounds. Toxicol Sci. 2006;93(2):223–41.
19. Spitsbergen JM, Kleeman JM, Peterson RE. Morphologic lesions and acute toxicity in rainbow trout (Salmo Gairdneri) treated with 2,3,7,8-tetrachlorodibenzo-p-dioxin. J Toxicol Environ Health. 1988;23(3):333–58.
20. Olson JR, McGarrigle BP, Tonucci DA, Schecter A, Eichelberger H. Developmental toxicity of

2,3,7,8-TCDD in the rat and hamster. Chemosphere. 1990;20:1117.

21. Zetterstrom R. Child health and environmental pollution in the Aral Sea region in Kazakhstan. Acta Paediatr Suppl. 1999;88(429):49–54.

22. Leijs MM, Koppe JG, Olie K, van Aalderen WM, Voogt P, Vulsma T, et al. Delayed initiation of breast development in girls with higher prenatal dioxin exposure; a longitudinal cohort study. Chemosphere. 2008;73(6):999–1004.

23. Leijs M, van der Linden L, Koppe JG, Olie K, van Aalderen W, ten Tusscher GW. The influence of perinatal and current dioxin and PCB exposure on reproductive parameters (sex-ratio, menstrual cycle characteristics, endometriosis, semen quality, and prematurity): a review. Biomonitoring. 2014;1(1):1–15.

24. Leijs M, van der Linden L, Koppe J, de Voogt P, Olie K, van Aalderen W, et al. The influence of perinatal and current dioxin and PCB exposure on puberty: a review. Biomonitoring. 2014;1(1):16–24.

25. World Health Organisation. Levels of PCBs, PCDDs and PCDFs in breast milk. Copenhagen: WHO; 1989.

26. World Health Organisation. Levels of PCBs, PCDDs and PCDFs in human milk. Bilthoven: WHO; 1996.

27. Goldman LR. Chemicals and children's environment: what we don't know about risks. Environ Health Perspect. 1998;106(Suppl 3):875–80.

28. Brouwer A, Longnecker MP, Birnbaum LS, Cogliano J, Kostyniak P, Moore J, et al. Characterization of potential endocrine-related health effects at low-dose levels of exposure to PCBs. Environ Health Perspect. 1999;107(Suppl 4):639–49.

29. Gore AC, Wu TJ, Oung T, Lee JB, Woller MJ. A novel mechanism for endocrine-disrupting effects of polychlorinated biphenyls: direct effects on gonadotropin-releasing hormone neurones. J Neuroendocrinol. 2002;14(10):814–23.

30. Nelson JA. Effects of dichlorodiphenyltrichloroethane (DDT) analogs and polychlorinated biphenyl (PCB) mixtures on 17beta-(3H)estradiol binding to rat uterine receptor. Biochem Pharmacol. 1974;23(2):447–51.

31. Hansen LG. Stepping backward to improve assessment of PCB congener toxicities. Environ Health Perspect. 1998;106(Suppl 1):171–89.

32. Leijs MM, ten Tusscher GW, Olie K, van Teunenbroek T, van Aalderen WM, de Voogt P, et al. Thyroid hormone metabolism and environmental chemical exposure. Environ Health. 2012;11 Suppl 1:S10.

33. Hsu ST, Ma CI, Hsu SK, SS W, Hsu NH, Yeh CC, et al. Discovery and epidemiology of PCB poisoning in Taiwan: a four-year followup. Environ Health Perspect. 1985;59:5–10.

34. Yoshimura T. Yusho in Japan. Ind Health. 2003;41(3):139–48.

35. Chen YC, Guo YL, Hsu CC, Rogan WJ. Cognitive development of Yu-Cheng ("oil disease") children prenatally exposed to heat-degraded PCBs. JAMA. 1992;268(22):3213–8.

36. Rogan WJ, Gladen BC, Hung KL, Koong SL, Shih LY, Taylor JS, et al. Congenital poisoning by polychlorinated biphenyls and their contaminants in Taiwan. Science. 1988;241(4863):334–6.

37. Bertazzi PA, Domenico A. Health consequences of the Seveso, Italy, accident. In: Schecter A, Gasiewicz TA, editors. Dioxins and health. 2nd ed. New York: John Wiley & Sons; 2003. p. 827–53.

38. Eskenazi B, Mocarelli P, Warner M, Samuels S, Vercellini P, Olive D, et al. Seveso Women's Health Study: a study of the effects of 2,3,7,8-tetrachlorodibenzo-p-dioxin on reproductive health. Chemosphere. 2000;40(9–11):1247–53.

39. Stellman SD, Stellman JM, Sommer JF Jr. Combat and herbicide exposures in Vietnam among a sample of American legionnaires. Environ Res. 1988;47(2):112–28.

40. Dutta SK, Mitra PS, Ghosh S, Zang S, Sonneborn D, Hertz-Picciotto I, et al. Differential gene expression and a functional analysis of PCB-exposed children: understanding disease and disorder development. Environ Int. 2012;40:143–54. doi:10.1016/j.envint.2011.07.008.

41. Darnerud PO, Eriksen GS, Johannesson T, Larsen PB, Viluksela M. Polybrominated diphenyl ethers: occurrence, dietary exposure, and toxicology. Environ Health Perspect. 2001;109(Suppl 1):49–68.

42. Siddiqi MA, Laessig RH, Reed KD. Polybrominated diphenyl ethers (PBDEs): new pollutants-old diseases. Clin Med Res. 2003;1(4):281–90.

43. Hooper K, McDonald TA. The PBDEs: an emerging environmental challenge and another reason for breast-milk monitoring programs. Environ Health Perspect. 2000;108(5):387–92.

44. Roper CS, Simpson AG, Madden S, Serex TL, Biesemeier JA. Absorption of [14C]-tetrabromodiphenyl ether (TeBDE) through human and rat skin in vitro. Drug Chem Toxicol. 2006;29(3):289–301.

45. Zhou T, Taylor MM, DeVito MJ, Crofton KM. Developmental exposure to brominated diphenyl ethers results in thyroid hormone disruption. Toxicol Sci. 2002;66(1):105–16.

46. Costa LG, Giordano G, Tagliaferri S, Caglieri A, Mutti A. Polybrominated diphenyl ether (PBDE) flame retardants: environmental contamination, human body burden and potential adverse health effects. Acta Biomed. 2008;79(3):172–83.

47. Darnerud PO. Toxic effects of brominated flame retardants in man and in wildlife. Environ Int. 2003;29(6):841–53.

48. Hamers T, Kamstra JH, Sonneveld E, Murk AJ, Kester MH, Andersson PL, et al. In vitro profiling of the endocrine-disrupting potency of brominated flame retardants. Toxicol Sci. 2006;92(1):157–73.

49. Legler J, Brouwer A. Are brominated flame retardants endocrine disruptors? Environ Int. 2003;29(6):879–85.

50. Abdelouahab N, Suvorov A, Pasquier JC, Langlois MF, Praud JP, Takser L. Thyroid disruption by low-dose BDE-47 in prenatally exposed lambs. Neonatology. 2009;96(2):120–4.

51. Hallgren S, Sinjari T, Hakansson H, Darnerud PO. Effects of polybrominated diphenyl ethers (PBDEs) and polychlorinated biphenyls (PCBs) on thyroid hormone and vitamin A levels in rats and mice. Arch Toxicol. 2001;75(4):200–8.

52. Stoker TE, Laws SC, Crofton KM, Hedge JM, Ferrell JM, Cooper RL. Assessment of DE-71, a commercial polybrominated diphenyl ether (PBDE) mixture, in the EDSP male and female pubertal protocols. Toxicol Sci. 2004;78(1):144–55.

53. van der Ven LT, van de Kuil T, Verhoef A, Leonards PE, Slob W, Canton RF, et al. A 28-day oral dose toxicity study enhanced to detect endocrine effects of a purified technical pentabromodiphenyl ether (pentaBDE) mixture in Wistar rats. Toxicology. 2008;245(1–2):109–22.

54. Zhang S, Bursian SJ, Martin PA, Chan HM, Tomy G, Palace VP, et al. Reproductive and developmental toxicity of a pentabrominated diphenyl ether mixture, DE-71(R), to ranch mink (Mustela vison) and hazard assessment for wild mink in the Great Lakes region. Toxicol Sci. 2009;110(1):107–16.

55. Zhou T, Ross DG, DeVito MJ, Crofton KM. Effects of short-term in vivo exposure to polybrominated diphenyl ethers on thyroid hormones and hepatic enzyme activities in weanling rats. Toxicol Sci. 2001;61(1):76–82.

56. Birnbaum LS, Staskal DF. Brominated flame retardants: cause for concern? Environ Health Perspect. 2004;112(1):9–17.

57. Branchi I, Capone F, Alleva E, Costa LG. Polybrominated diphenyl ethers: neurobehavioral effects following developmental exposure. Neurotoxicology. 2003;24(3):449–62.

58. Costa LG, Giordano G. Developmental neurotoxicity of polybrominated diphenyl ether (PBDE) flame retardants. Neurotoxicology. 2007;28(6):1047–67.

59. Hites RA, Foran JA, Schwager SJ, Knuth BA, Hamilton MC, Carpenter DO. Global assessment of polybrominated diphenyl ethers in farmed and wild salmon. Environ Sci Technol. 2004;38(19):4945–9.

60. ten Tusscher GW. Later childhood effects of perinatal exposure to background levels of dioxins in the Netherlands. Amsterdam: University of Amsterdam; 2002.

61. Fowles JR, Fairbrother A, Baecher-Steppan L, Kerkvliet NI. Immunologic and endocrine effects of the flame-retardant pentabromodiphenyl ether (DE-71) in C57BL/6J mice. Toxicology. 1994;86(1–2):49–61.

62. Leijs MM, Koppe JG, Olie K, van Aalderen WM, de VP, ten Tusscher GW. Effects of dioxins, PCBs, and PBDEs on immunology and hematology in adolescents. Environ Sci Technol. 2009;43(20):7946–51.

63. Zhou J, Chen DJ, Liao QP, Yu YH. Impact of PBDE-209 exposure during pregnancy and lactation on immune function of offspring rats. J South Med Uni (Nan Fang Yi Ke Da Xue Xue Bao). 2006;26(6):738–41.

64. Leijs MM. Toxic effects of dioxins, PCBs and PBDEs in adolescents. Amsterdam: Ph.D Thesis University of Amsterdam; 2010.

65. Chanda JJ, Anderson HA, Glamb RW, Lomatch DL, Wolff MS, Voorhees JJ, et al. Cutaneous effects of exposure to polybrominated biphenyls (PBBs): the Michigan PBB incident. Environ Res. 1982;29(1):97–108.

66. Ouw HK, Simpson GR, Siyali DS. Use and Health Effects of Aroclor 1242, a Polychlorinated Biphenyl, in an Electrical Industry. Archives of Environmental Health: An International Journal. 1976; 31 (4): 189–194.

67. Fischbein A, Wolff MS, Lilis R, Thornton J, Selikoff IJ. Clinical findings among PCB-exposed capacitor manufacturing workers. Ann NY Acad Sci. 1979; 320: 703–15.

68. Fischbein A, Rizzo JN, Solomon SJ, Wolff MS. Oculodermatological findings in workers with occupational exposure to polychlorinated biphenyls (PCBs). Br J Ind Med 1985 Jun;42(6):426-30.

69. Saurat JH, Kaya G, Saxer-Sekulic N, Pardo B, Becker M, Fontao L, et al. The cutaneous lesions of dioxin exposure: lessons from the poisoning of Victor Yushchenko. Toxicol Sci. 2012;125(1):310–7.

70. Geusau A, Abraham K, Geissler K, Sator MO, Stingl G, Tschachler E. Severe 2,3,7,8-tetrachlorodibenzo-p-dioxin (TCDD) intoxication: clinical and laboratory effects. Environ Health Perspect. 2001;109(8):865–9.

71. Geusau A, Tschachler E, Meixner M, Papke O, Stingl G, McLachlan M. Cutaneous elimination of 2,3,7,8-tetrachlorodibenzo-p-dioxin. Br J Dermatol. 2001;145(6):938–43.

72. Cheng WN, Coenraads PJ, Hao ZH, Liu GF. A health survey of workers in the pentachlorphenol section of a chemical manufacturing plant. 1993; 24(1): 81–92).

73. Piagitelli L, Marlow D, Fingerhut M, Steenland K, Sweeney MH. A retrospective job exposure matrix for estimating exposure to 2,3,8,8-tetrachlorodibenzo-p-dioxin. Am J Ind Med. 2000; 38(1):28–39.

74. Caramaschi F, del Corno G, Favaretti C, Giambelluca SE, Montesarchio E, Fara GM. Chloracne following environmental contamination by TCDD in Seveso, Italy. Int J Epidemiol. 1981; 10(2): 135–43.

75. Sorg O, Zennegg M, Schmid P, Fedosyuk R, Valikhnovskyi R, Gaide O, et al. 2,3,7,8-tetrachlorodibenzo-p-dioxin (TCDD) poisoning in Victor Yushchenko: identification and measurement of TCDD metabolites. Lancet. 2009;374(9696):1179–85.

76. Michielsen CPPC, Bloksma N, Ultee A, van Mil F, Vos JG. Hexachlorobenzene-induced immunomodulation in skin and lung lesions: a comparison between brown Norway, Lewis, and Wistar rats. Toxicol Appl Pharmacol. 1997;144:12–26.

77. Von Bettmann S. Chlorakne eine besondere Form von professioneller Hauterkrankung. Dtsch Med Wochenschr. 1901;27:437.

78. Herxheimer K. Ueber chlorakne. Munch Med Wochenschr. 1899;46:278.

79. Kimmig J, Schulz KH. Occupational acne (so-called chloracne) due to chlorinated aromatic cyclic ethers. Dermatologica. 1957;115(4):540–6.

80. Schulz KH. Clinical & experimental studies on the etiology of chloracne. Arch Klin Exp Dermatol. 1957;206:589–96.

81. Ju Q, Zouboulis CC, Xia L. Environmental pollution and acne: chloracne. Dermatoendocrinol. 2009;1(3):125–8.

82. Panteleyev AA, Bickers DR. Dioxin-induced chloracne--reconstructing the cellular and molecular mechanisms of a classic environmental disease. Exp Dermatol. 2006;15(9):705–30.

83. Poland A, Knutson JC. 2,3,7,8-tetrachlorodibenzo-p-dioxin and related halogenated aromatic hydrocarbons: examination of the mechanism of toxicity. Annu Rev Pharmacol Toxicol. 1982;22:517–54.

84. Tindall JP. Chloracne and chloracnegens. J Am Acad Dermatol. 1985;13(4):539–58.

85. Zugerman C. Chloracne. Clinical manifestations and etiology. Dermatol Clin. 1990;8(1):209–13.

86. Jensen NE, Sneddon IB, Walker AE. Tetrachlorobenzodioxin and chloracne. Trans St Johns Hosp Dermatol Soc. 1972;58(2):172–7.

87. McDonagh AJ. Chloracne-study of an outbreak with new clinical observations. Clin Exp Dermatol. 1993;18(6):523–5.

88. Reggiani G, Bruppacher R. Symptoms, signs and findings in humans exposed to PCBs and their derivatives. Environ Health Perspect. 1985;60:225–32.

89. Zober A, Ott MG, Messerer P. Morbidity follow up study of BASF employees exposed to 2,3,7,8-tetrachlorodibenzo-p-dioxin (TCDD) after a 1953 chemical reactor incident. Occup Environ Med. 1994;51(7):479–86.

90. McKee M. The poisoning of Victor Yushchenko. Lancet. 2009;374:1131–2.

91. May G. Chloracne from the accidental production of tetrachlorodibenzodioxin. Br J Ind Med. 1973;30(3):276–83.

92. Baccarelli A, Pesatori AC, Consonni D, Mocarelli P, Patterson DG Jr, Caporaso NE, et al. Health status and plasma dioxin levels in chloracne cases 20 years after the Seveso, Italy accident. Br J Dermatol. 2005;152(3):459–65.

93. Coenraads PJ, Brouwer A, Olie K, Tang N. Chloracne. Some recent issues. Dermatol Clin. 1994;12(3):569–76.

94. Liu J, Zhang CM, Coenraads PJ, Ji ZY, Chen X, Dong L, Ma XM, Han W, Tang NJ. Abnormal expression of MAPK, EGFR, CK17 and TGk in the skin lesions of chloracne patients exposed to dioxins. 2011 201(3):230-234.

95. Schafer M, Willrodt AH, Kurinna S, Link AS, Farwanah H, Geusau A, et al. Activation of Nrf2 in keratinocytes causes chloracne (MADISH)-like skin disease in mice. EMBO Mol Med. 2014;6(4):442–57.

96. Moses M, Prioleau PG. Cutaneous histologic findings in chemical workers with and without chloracne with past exposure to 2,3,7,8-tetrachlorodibenzo-p-dioxin. J Am Acad Dermatol. 1985;12(3):497–506.

97. Loertscher JA, Sadek CS, len-Hoffmann BL. Treatment of normal human keratinocytes with 2,3,7,8-tetrachlorodibenzo-p-dioxin causes a reduction in cell number, but no increase in apoptosis. Toxicol Appl Pharmacol. 2001;175(2):114–20.

98. Birnbaum LS. The mechanism of dioxin toxicity: relationship to risk assessment. Environ Health Perspect. 1994;102(Suppl 9):157–67.

99. Plewig G, Albrecht G, Henz BM, Meigel W, Schopf E, Stadler R. Systemic treatment of acne with isotretinoin: current status. Hautarzt. 1997;48(12):881–5.

100. Yip J, Peppall L, Gawkrodger DJ, Cunliffe WJ. Light cautery and EMLA in the treatment of chloracne lesions. Br J Dermatol. 1993;128(3):313–6.

101. Geusau A, Tschachler E, Meixner M, Sandermann S, Papke O, Wolf C, et al. Olestra increases faecal excretion of 2,3,7,8-tetrachlorodibenzo-p-dioxin. Lancet. 1999;354(9186):1266–7.

102. Jandacek RJ, Anderson N, Liu M, Zheng S, Yang Q, Tso P. Effects of yo-yo diet, caloric restriction, and olestra on tissue distribution of hexachlorobenzene. Am J Physiol Gastrointest Liver Physiol. 2005;288(2):G292–9.

103. Geusau A, Schmaldienst S, Derfler K, Papke O, Abraham K. Severe 2,3,7,8-tetrachlorodibenzo-p-dioxin (TCDD) intoxication: kinetics and trials to enhance elimination in two patients. Arch Toxicol. 2002;76(5–6):316–25.

104. Caramaschi F, del CG, Favaretti C, Giambelluca SE, Montesarchio E, Fara GM. Chloracne following environmental contamination by TCDD in Seveso, Italy. Int J Epidemiol. 1981;10(2):135–43.

105. Cook RR. Dioxin, chloracne, and soft tissue sarcoma. Lancet. 1981;1(8220 Pt 1):618–9.

106. Reggiani G. Acute human exposure to TCDD in Seveso, Italy. J Toxicol Environ Health. 1980;6(1):27–43.

107. Leijs MM, Amann P, Werthan A, et al. Skin manifestations in German workers with high occupational PCB exposure. Organohalogen Compd. 2014;76:1577–80.

108. Kawasaki G, Yoshitomi I, Yanamoto S, Yamada S, Mizuno A, Umeda M. Pigmentation of the oral mucosa by PCB poisoning in Yusho patients. Arch Oral Biol. 2013;58(9):1260–4.

109. Luecke S, Backlund M, Jux B, Esser C, Krutmann J, Rannug A. The aryl hydrocarbon receptor (AHR), a novel regulator of human melanogenesis. Pigment Cell Melanoma Res. 2010;23(6):828–33.

110. Zodrow JM, Tanguay RL. 2,3,7,8-tetrachlorodibenz o-p-dioxin inhibits zebrafish caudal fin regeneration. Toxicol Sci. 2003;76(1):151–61.

111. Shertzer HG, Nebert DW, Puga A, Ary M, Sonntag D, Dixon K, et al. Dioxin causes a sustained oxidative

stress response in the mouse. Biochem Biophys Res Commun. 1998;253(1):44–8.

112. Tritscher AM, Seacat AM, Yager JD, Groopman JD, Miller BD, Bell D, et al. Increased oxidative DNA damage in livers of 2,3,7,8-tetrachlorodibenzo-p-dioxin treated intact but not ovariectomized rats. Cancer Lett. 1996;98(2):219–25.

113. Osborne R, Greenlee WF. 2,3,7,8-Tetrachlorodibenzo-p-dioxin (TCDD) enhances terminal differentiation of cultured human epidermal cells. Toxicol Appl Pharmacol. 1985;77(3):434–43.

114. Merk H, Bolsen K, Lissner R, Goerz G. Hexachlorobenzene alteration of benzo[a]pyrene metabolism in porphyric and non-porphyric rats. IARC Sci Publ. 1986;77:461–3.

115. Geusau A, Jurecka W, Nahavandi H, Schmidt JB, Stingl G, Tschachler E. Punctate keratoderma-like lesions on the palms and soles in a patient with chloracne: a new clinical manifestation of dioxin intoxication? Br J Dermatol. 2000;143(5):1067–71.

116. Coenraads PJ, Olie K, Tang NJ. Blood lipid concentrations of dioxins and dibenzofurans causing chloracne. Br J Dermatol. 1999;141(4):694–7.

117. Ono R, Kagawa Y, Takahashi Y, Akagi M, Kamei C. Effect of 2,3,7,8-tetrachlorodibenzo-p-dioxin on scratching behavior in mice. Int Immunopharmacol. 2010;10(3):304–7.

118. Fischbein A, Rizzo JN, Solomon SJ, Wolff MS. Oculodermatological findings in workers with occupational exposure to polychlorinated biphenyls (PCBs). Br J Ind Med. 1985;42(6):426–30.

119. Hicks N, Zack M, Caldwell GG, McKinley TW. Lifestyle factors among patients with melanoma. South Med J. 1985;78(8):903–8.

120. Loomis D, Browning SR, Schenck AP, Gregory E, Savitz DA. Cancer mortality among electric utility workers exposed to polychlorinated biphenyls. Occup Environ Med. 1997;54(10):720–8.

121. Jang MS, Jang JG, Han SH, Park JB, Kang DY, Kim ST, et al. Clinicopathological features of mycosis fungoides in patients exposed to agent orange during the Vietnam War. J Dermatol. 2013;40(8):606–12.

122. Akhtar FZ, Garabrant DH, Ketchum NS, Michalek JE. Cancer in US air force veterans of the Vietnam war. J Occup Environ Med. 2004;46(2):123–36.

123. Gallagher RP, Macarthur AC, Lee TK, Weber JP, Leblanc A, Mark EJ, et al. Plasma levels of polychlorinated biphenyls and risk of cutaneous malignant melanoma: a preliminary study. Int J Cancer. 2011;128(8):1872–80.

124. Silberberg I, Baer RL, Rosenthal SA. The role of Langerhans cells in allergic contact hypersensitivity. A review of findings in man and guinea pigs. J Invest Dermatol. 1976;66(4):210–7.

125. Bos JD. Huid en afweer, inaugural oration. Int Rev Cytol. 1991. Amsterdam, Bohn, Stafleu, van Loghum.

126. ten Tusscher GW, Leijs MM, Olie K, Ilsen A, Vulsma T, Koppe JG. Findings on prenatal, lactational and later childhood exposure to dioxins and dioxin-like compounds: a review of the Amsterdam-Zaandam cohort 1987–2005. AIMS Environ Sci. 2015;2(1):1–20.

127. Serdar B, Leblanc WG, Norris JM, Dickinson LM. Potential effects of polychlorinated biphenyls (PCBs) and selected organochlorine pesticides (OCPs) on immune cells and blood chemistry measures: a cross-sectional assessment of the NHANES 2003–2004 data. Environ Health. 2014;13:114.

128. Boule LA, Winans B, Lawrence BP. Effects of developmental activation of the AhR on CD4 + T-cell responses to influenza virus infection in adult mice. Environ Health Perspect. 2014;122:1201–8.

129. Geusau A, Khorchide M, Mildner M, Pammer J, Eckhart L, Tschachler E. 2,3,7,8-Tetrachlorodibenzo-p-dioxin impairs differentiation of normal human epidermal keratinocytes in a skin equivalent model. J Invest Dermatol. 2005;124(1):275–7.

Zeitfracht Medien GmbH
Ferdinand-Jühlke-Straße 7
99095 Erfurt, Deutschland
produktsicherheit@kolibri360.de